Teacher's guide to Book 8S

Contents

CAMBRIDGE UNIVERSITY PRESS

CAMBRIDGE UNIVERSITY PRESS
Cambridge, New York, Melbourne, Madrid, Cape Town, Singapore, São Paulo

Cambridge University Press
The Edinburgh Building, Cambridge CB2 2RU, UK

www.cambridge.org
Information on this title: www.cambridge.org/9780521538053

© The School Mathematics Project 2003

First published 2003
5th printing 2006

Printed in the United Kingdom at the University Press, Cambridge

A catalogue record for this publication is available from the British Library

ISBN-13 978-0-521-53805- 3 paperback
ISBN-10 0-521-53805- X paperback

Typesetting and technical illustrations by The School Mathematics Project
Cover image © ImageState Ltd
Cover design by Angela Ashton

The maps on pages 109 and 110 are based on Ordnance Survey mapping with
the permission of the Controller of Her Majesty's Stationery Office.
© Crown copyright, Licence No. 100001679.

The following people contributed to the writing of the SMP Interact key stage 3 materials.

Ben Alldred	Ian Edney	John Ling	Susan Shilton
Juliette Baldwin	Steve Feller	Carole Martin	Caroline Starkey
Simon Baxter	Rose Flower	Peter Moody	Liz Stewart
Gill Beeney	John Gardiner	Lorna Mulhern	Pam Turner
Roger Beeney	Bob Hartman	Mary Pardoe	Biff Vernon
Roger Bentote	Spencer Instone	Peter Ransom	Jo Waddingham
Sue Briggs	Liz Jackson	Paul Scruton	Nigel Webb
David Cassell	Pamela Leon	Richard Sharpe	Heather West

Others, too numerous to mention individually, gave valuable advice, particularly by commenting on and trialling draft materials.

Editorial team	Project administrator	Design	Project support
David Cassell	Ann White	Pamela Alford	Carol Cole
Spencer Instone		Melanie Bull	Pam Keetch
John Ling		Nicky Lake	Jane Seaton
Paul Scruton		Tiffany Passmore	Cathy Syred
Susan Shilton		Martin Smith	
Caroline Starkey			
Heather West			

Special thanks go to Colin Goldsmith.

Introduction

Teaching approaches

SMP Interact sets out to help teachers use a variety of teaching approaches in order to stimulate pupils and foster their understanding and enjoyment of mathematics.

A central place is given to discussion and other interactive work. In this respect and others the material supports the methodology of the *Framework for teaching mathematics*. Questions that promote effective discussion and activities well suited to group work occur throughout the material.

Some activities, mostly where a new idea or technique is introduced, are described only in the teacher's guide. (These are indicated in the pupils' book by a solid marginal strip – see below.)

Materials

There are three series in key stage 3: books 7T–9T cover up to national curriculum level 5; 7S–9S go up to level 6; 7C–9C go up to level 7, though schools have successfully prepared pupils for level 8 with them, drawing lightly on extra topics from early in the *SMP Interact* GCSE course.

Pupils' books

Each unit of work begins with a statement of learning objectives and most units end with questions for self-assessment.

  Teacher-led activities that are described in the teacher's guide are denoted by a solid marginal strip in both the pupil's book and the teacher's guide.

Some other activities that are expected to need teacher support are marked by a broken strip.

Where the writers have particular classroom organisation in mind (for example working in pairs or groups), this is stated in the pupils' book.

Resource sheets

Resource sheets, some essential and some optional, are linked to some activities in the books.

Practice booklets

For each book there is a practice booklet containing further questions unit by unit. These booklets are particularly suitable for homework.

Teacher's guides

For each unit, there is usually an overview, details of any essential or optional equipment, including resource sheets, and the practice booklet page references, followed by guidance that includes detailed descriptions of teacher-led activities, advice on difficult ideas and comments from teachers who trialled the material.

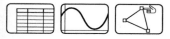

There is scope to use computers and graphic calculators throughout the material. These symbols mark specific opportunities to use a spreadsheet, graph plotter and dynamic geometry software respectively.

Answers to questions in the pupils' book and the practice booklet follow the guidance. For reasons of economy answers to resource sheets that pupils write on are not always given in the teacher's guide; they can of course be written on a spare copy of the sheet.

Assessment

Unit by unit assessment tests are available both as hard copy and as editable files on CD (details are at www.smpmaths.org.uk). The practice booklets are also suitable as an assessment resource.

Oral and mental starters

An oral or mental starter can be used for a number of purposes.

- It can **introduce the main topic**, and many of the teacher-led activities described in this guide can be used in this way.

- It can also be an effective way of **revising skills that are needed for the main topic** and can prevent the subsequent lesson 'sagging' when those skills falter. For example, a 'matching' starter of the kind described on the next page can be used to revise percentages and multipliers before work on percentage change (unit 15).

- Alternatively a starter can be used to **revise skills that are unrelated to the main lesson**. Some questions and activities in the pupils' book can be adapted for later use as starters. For example, question A5 on page 86 can be used with integers, decimals, fractions or algebraic expressions.

Starter formats

The formats described below have been found very effective and can be adapted to different topics.

Small whiteboards ('show me boards') and markers are invaluable for many types of starter, for example *True or false?*, *What am I?* and *Odd one out*. Pupils write their response on their board and hold it up, giving you instant feedback on the whole class.

True or false?

You say or write a statement (such as '1 inch ≈ 2.5 cm', '9 is the square root of 3' or '$2^3 = 6$') and pupils decide if it is true or false.

Spider diagram

You write a whole number, fraction, decimal, percentage, word, algebraic expression … in a circle on the board. This is the spider's 'body'. The 'legs' and 'feet' can be labelled in a variety of ways.

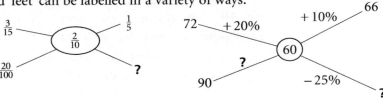

Today's number is …

Write a number on the board and put a ring round it. It could be a whole number (including negatives), decimal or fraction. Pupils make up calculations with that number as the result. Calculations can be restricted to a particular type. A variation of this format is *Today's expression is …*

What am I?

For example, 'I am a quadrilateral. I have four equal sides. I have rotation symmetry of order 2. What am I?'

Ordering

Pupils put a set of numbers such as 2^3, 3^2, 2^5, 5^2 in order of size.

Matching

Pupils match up a set (such as this one) into pairs or larger groups.

| 20% | 2% | 0.02 | 0.22 | 0.2 | 22% |

Target number

Pupils are given a set of numbers or choose them themselves.
A 'target' number is chosen. Pupils try to make the target number using the given numbers and any of the four operations. (Not all numbers need be used but each may be used only once.)

For example, make 243 from these.

    4   5   6   3   10   25

Bingo

Pupils choose numbers from a given set and fill in their bingo 'card' with them. The card can hold as many numbers as you choose.

In this example, numbers have been chosen from a given set of fifteen. You could 'call':

- $2400 \div 600$

- $4.8 \div 0.6$

	2		6
		9	
	12		8

Counting stick

A counting stick is marked in equal intervals, usually in two contrasting colours. You can show pupils a starting point, tell them the value of one interval and then ask them to identify numbers you point to. For example, you could tell them one interval on this stick is 0.01 .

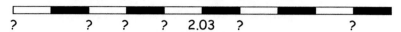

?   ?   ?   ?   2.03   ?   ?

Alternatively, give the values of two key points (Blu-tack labels on) and indicate points at random, each time asking for the value.

Array

This is a set of numbers, fractions, expressions ... arranged in a grid or just in a list, on which you base questions. For example,

- find three prime numbers
- find two numbers with 15 as their product
- find three numbers with 25 as their sum
- find two numbers with a difference of 3

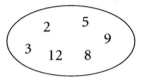

Odd one out

Pupils identify the odd one out from a set, such as $2(2a + b)$, $4a + 2b$, $3a + 2b + a - b$, $7a + 5b - 3a - 3b$.

Two-way property table

The headings of a two-way table like this one are shown on the board and pupils say what could go in the cells. It is often best to say that all the entries should be different.

	A factor of 6	A cube number
A multiple of 3		
A square number		

Topics for starters

Below, arranged by some broad topic areas, are suggestions for other ways the starter formats can be used.

Integer calculations

Array Use a set of positive and negative numbers.

True or false For example, $40 \times 50 = 200$, $^-8 \times {}^-5 = 40$

Number relationships

Bingo Use 'calls' like:

- a prime number that is a factor of 25
- 2 cubed times 3
- the square of 4

What am I? I am a square number with two digits. The sum of my digits is 13. What am I?

Algebraic manipulation

True or false? For example, $3(a + 2b) = 3a + 6b$

Spider diagram An expression such as $10 - h^2$ can be put on the body and the values of the expression for different values of h can be put on the legs.

Today's expression is … Pupils have to find combinations of expressions that are equivalent to a given expression. For example 'Today's expression is $6a + 3b$' could lead to $3(2a + b)$, $5a + 8b + a - 5b$ and so on.

Matching A set of expressions has to be sorted into groups or pairs of equivalent expressions.

Fractions and ratio

True or false? For example, '$2:3$ is the same ratio as $4:5$', '$4 \div \frac{1}{2} = 2$'

Ordering For example, a set of paint recipes that are sorted by strength of colour, a set of fractions such as $\frac{10}{3}$, $1\frac{1}{3}$, $\frac{5}{3}$, $2\frac{2}{3}$

Odd one out Show a set of ratios such as $2:1$, $4:2$, $5:10$, $6:3$.

Matching Pupils match a set of improper fractions with a set of mixed numbers.

Percentages

Matching A set of increases, decreases and multipliers can be sorted into equivalent pairs; fractions can be matched with percentages.

Today's number is … Pupils have to devise percentage calculations that result in a given number.

Ordering For example, 0.25, $\frac{1}{5}$, 23%, $\frac{1}{8}$, 0.4, 38%

Bingo Questions called are 'What is 10% of 90?', 'Increase 8 by 25%' and so on.

Decimals and measures

Array Use decimals and ask questions such as 'Find two numbers that have a sum of …', 'Write down the numbers that are between 4.4 and 4.5'.

True or false? For example, $1.3 \times 10 = 1.30$, 1.3 litres = 1300 ml, $1 \, cm^2 = 10 \, mm^2$

Odd one out Use a set such as 0.3×0.8, 0.03×8, 0.6×0.4, 0.06×40.

Ordering For example, a set of rectangles that are sorted by area or perimeter, a set of lengths such as 500 m, 0.6 km, 0.58 km, 1300 m

Spatial visualisation

Spider diagram A shape such as a square can be put on the body and the shape's properties can be put on the feet.

What am I? 'I have three sides and one line of symmetry. What am I?' 'I have 4 equal sides and 4 lines of symmetry. What am I?'

① Into the bath (p 4)

This involves interpretation of the overall features of graphs, particularly the shapes of curves, rather than focusing on the values for particular points.

Optional

Sheet 191 for use with question 2

Practice booklet pages 3 and 4

◊ Start by asking the pupils to look at the graph of the water level in Peter's bath. Emphasise that the graph shows the **water level, not the volume** of water in the bath. Bring out what each section of the graph implies is happening (see below) but defer discussion of what happens after 30 minutes since question A1 covers this.

'They did enjoy writing and sharing their stories.'

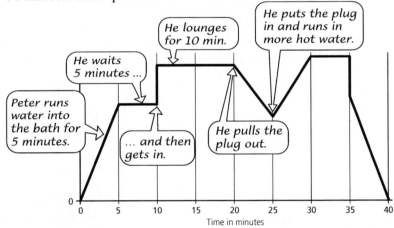

He waits 5 minutes ...

He lounges for 10 min.

He puts the plug in and runs in more hot water.

Peter runs water into the bath for 5 minutes.

... and then gets in.

He pulls the plug out.

Time in minutes

2 Some pupils may be helped if they have a copy of sheet 191. They can label events using bubbles as above or can mark key points and refer to them in their stories.

1 (a) Peter is lounging (the water level is constant).

(b) He gets out and pulls out the plug.

(c) The water is running out.

2 The pupil's stories corresponding to the three graphs. In the following timings times like 17 minutes are only approximate.

Nikki's bath

0 to 5 min	runs water
at 5 min	gets in
5 to 10 min	lounges
10 to 15 min	runs more water
15 to 17 min	lounges
17 to 20 min	lets out some water
20 to 30 min	lounges
at 30 min	gets out and pulls out plug
at 35 min	bath is empty

Mike's bath

0 to 5 min	runs water
at 5 min	gets in and immediately pulls out plug
5 to 7 min	water drains out
7 to 13 min	puts plug back in and runs more water
13 to 20 min	lounges
20 to 23 min	lets out some water
23 to 25 min	puts plug in and runs more water
25 to 30 min	lounges
30 to 35 min	lets out water
at 35 min	gets out
at $37\frac{1}{2}$ min	bath is empty

Chris's bath

0 to 5 min	runs water
at 5 min	gets in and almost immediately out!
6 to 10 min	runs more water (probably cold!)
at 10 min	gets back in
10 to 20 min	lounges but water is slowly running out
at 20 min	puts the plug in properly!
20 to 23 min	lounges
23 to 24 min	runs more water
24 to 30 min	lounges
at 30 min	gets out and pulls the plug out
at 35 min	bath is empty

3

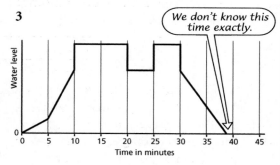

We don't know this time exactly.

4 The pupil's sketch graph and another pupil's story

Practice booklet (p 3)

1 (a) It stays at the 2nd floor for a short time, then goes down to the 1st floor, waits for 20 seconds and then goes down to the basement.

(b) The lift starts to go directly up to the 5th floor.

(c) About 10 seconds

(d) 2 minutes

(e) 1st floor

(f) About 2 minutes 20 seconds

2

0 to 5 min	runs one tap
5 to 10 min	runs other tap as well
10 min	gets in and immediately out
10 to 12 min	plug is out
12 to 20 min	runs in more water
at 20 min	gets in again
20 to 30 min	lounges
30 to 32 min	plug is out
32 to 35 min	runs in more water
35 to 40 min	lounges
at 40 min	gets out, takes out plug
40 to 48 min	water runs out

3 (a) 9 a.m.   (b) 8 p.m.

(c) About 5 p.m.   (d) More at 4 p.m.

(e) More coming in

(f) More leaving

2 Ratio

Essential	**Optional**
Sheet 209	Lime juice (or similar) cordial, water, plastic cups
	Linking cubes (at least two colours)
	Sheet 210
Practice booklet pages 5 to 9	

Ⓐ **Recipes** (p 6)

◊ A practical activity that some teachers and pupils have tried and found useful is a 'taste test'. It involves using the language of ratio to describe mixtures.

> Optional: lime juice (or similar) cordial, water and plastic cups.

'Orange juice takes some time to organise but is worthwhile.'

Pupils make up a variety of drinks using lime cordial and water in different ratios. They then put the cups out in a random order (after devising a way of telling for themselves which is which!) and other pupils try to put the drinks in order from strongest to weakest by tasting. They could start with simple ratios (1 measure of lime to 2 measures of water, 1 measure of lime to 3 measures of water, 1 measure of lime to 4 measures of water, …) and move on to ratios such as 2 measures of lime to 3 measures of water if appropriate.

The testing is unlikely to be conclusive! The activity serves its purpose if it gets pupils using phrases like '1 to 2'. Emphasise that order is important.

◊ In discussing the recipe for light orange paint, some pupils may think that adding the same amount of red and yellow will make the colour the same. Suggest this and encourage pupils to prove you wrong.

Ask questions such as:

- How much of each colour would you need to make three times as much light orange?
- How much yellow paint would you need to mix with 4 tins of red?
- How much yellow paint would you need to mix with 2 litres of red? How much light orange paint would this make?

   What about 2 thimblefuls of red paint?

- How much red paint would you need to mix with 10 tins of yellow?

◊ In discussing the different shades of orange, the idea is not for the pupils to hit upon the 'right' answer but for you to see whether or not the answers are plausible and show some understanding of the idea of ratio.

B **Ratios** (p 8)

The context is earrings and a set of linking cubes is useful to model this.

> Optional: a set of linking cubes

Your discussion should introduce the concept of ratio as a way of describing proportions written as, say, '2 to 1'.

◊ You could make in advance a set of '2 blue, 1 yellow' blocks (to represent the earrings). Help pupils to see that, even without counting them, we know it is true that there are 2 blues for every 1 yellow, so the ratio of blues to yellows in the set is 2 to 1. Emphasise that the order is important.

◊ When this idea has been understood, go on to the reverse of it. Make a mixture of, say, 12 blue and 4 yellow and ask if it can be made into simple identical blocks (earrings). Show that the ratio of blues to yellows can be written 3 to 1.

C **Strong, stronger, strongest** (p 9)

'This went well – noisy – but well.'

This game involves comparing ratios, where the numbers are relatively straightforward.

> Sheet 209 (lemon juice and water cards)

Darker, lighter (p 10)

This activity involves comparing and ordering ratios written in the form $a:b$.

> Optional:
> The recipes are printed on sheet 210 for cutting up into cards.

◊ The recipes in order darker to lighter are:

    B, D (6:2 = 3:1)

    A, G, I (2:1 = 6:3 = 4:2)

    F, L (3:2 = 6:4)

    E, K, M (1:1 = 3:3 = 2:2)

    O (2:3)

    C, N, P (1:2 = 2:4 = 3:6)

    H, J (2:6 = 1:3)

E Working with ratios (p 11)

F Sharing in a given ratio (p 13)

G Comparing ratios (p 14)

◊ Pupils can work in groups to try to decide which is the darker colour each time.

◊ You could discuss alternative methods for the green and orange shades.

For example:

Green shades

Method 1		Method 2	
Grass	Leaf	Grass	Leaf
5:3	3:2	5:3	3:2
(× 2 × 2)	(× 3 × 3)	(× 3 × 3)	(× 5 × 5)
10:6	9:6	15:9	15:10

Some pupils will take a bit of time to see that both methods lead to the same conclusion, that the grass green is darker (contains proportionally more blue).

◊ You can deal with the idea of reducing each ratio to the form $1:n$ or $n:1$ in your initial discussion or wait till pupils meet it for themselves in question G8.

A Recipes (p 6)

A1 (a) 6 tins   (b) 12 tins

A2

	Blue	Yellow
2 times the recipe	6 tins	**4 tins**
3 times the recipe	**9 tins**	6 tins
5 times the recipe	**15 tins**	**10 tins**
10 times the recipe	**30 tins**	**20 tins**

A3 (a) 12 litres

     (b) (i) 8 litres   (ii) 10 litres

     (c) 5 litres

A4 (a) 8 litres

     (b) (i) 20 litres   (ii) 25 litres

A5 (a) 18 cups   (b) 28 litres

A6

Litres of blue	Litres of yellow	Litres of light green
4	**10**	**14**
6	**15**	**21**
8	**20**	**28**
10	**25**	**35**

A7

Litres of red	Litres of blue	Litres of light purple
3	**2**	**5**
6	**4**	**10**
9	6	**15**
12	8	**20**
18	12	**30**

B Ratios (p 8)

B1 (a) 3 to 1   (b) 1 to 2

     (c) 1 to 3   (d) 4 to 1

B2 (a) No   (b) Yes

     (c) 1 to 2   (d) 2 to 1

B3 (a) E, 2 to 1

     (b) (i) D, 3 to 1   (ii) C, 4 to 1

         (iii) B, 2 to 3   (iv) F, 3 to 2

         (v) A, 1 to 3

E Working with ratios (p 11)

E1 (a) 15 litres   (b) 10 litres

E2 (a) 15 litres   (b) 4 litres

E3 (a) 1:2   (b) 12   (c) 20

E4 4:1

E5 4:3

E6 (a) 4:3   (b) 3:4

E7 (a) 3:1   (b) 3:2   (c) 2:3   (d) 4:7

E8 $\underline{1:5} = 4:20 = 20:100 = 2:10$
$4:6 = \underline{2:3} = 10:15$
$\underline{3:1} = 9:3 = 12:4$
$\underline{3:2} = 12:8$

E9 10 litres blue, 20 litres yellow

E10 5 litres red, 15 litres white

E11 (a) 15 litres

     (b) 5 litres black, 25 litres white

     (c) 4 litres black, 10 litres white

     (d) 30 litres

     (e) 15 litres black, 25 litres white

F Sharing in a given ratio (p 13)

F1 Stuart £8, Shula £4

F2 Dawn £15, Eve £5

F3 (a) Jake £5, Poppy £10

     (b) $\frac{1}{3}$

F4 (a) Martin £5, Priya £15

     (b) $\frac{1}{4}$

     (c) $\frac{3}{4}$

F5 Eric £9, Betty £12

F6 (a) £16, £4   (b) £24, £36

     (c) £15, £9   (d) £25, £20

     (e) £7.50, £5   (f) £4.50, £1.50

     (g) £10, £7.50   (h) 80p, £1

F7 £80

F8 (a) James £750, Sarah £1250

(b) The answers have been rounded to the nearest penny.

1 year: J £800, S £1200
2 years: J £833.33, S £1166.67
3 years: J £857.14, S £1142.86
4 years: J £875, S £1125, etc.

(c) James's share goes up and Sarah's goes down as grandmother lives longer.

F9 (a) £2 (b) £6

F10 Xavier £20, Yasmin £60, Zak £120

F11 Carol £8, Doug £12, Eva £12

F12 (a) £4, £6, £8

(b) £11, £16.50, £38.50

Ⓖ **Comparing ratios** (p 14)

G1 (a) $3:5 = \textbf{6}:10$ (b) $3:2 = 9:\textbf{6}$
(c) $7:10 = \textbf{21}:30$ (d) $4:9 = 20:\textbf{45}$

G2 Pale blue

G3 Psychedelic purple

G4 Thundercloud grey

G5 Simon's drink

G6 Jasmine's drink

G7 (a) $10:5 = \textbf{2}:1$ (b) $7:4 = \textbf{1.75}:1$
(c) $5:12 = 1:\textbf{2.4}$ (d) $0.5:3 = 1:\textbf{6}$

G8 (a) $1:3.4$

(b) Ken's drink is stronger as there is slightly less water for 1 part of juice.

G9 (a) Calypso punch $2.75:1$
Tangy punch $2.33:1$ (to 2 d.p.)

(b) Calypso punch

G10 (a) Sugar lilac $1:2.1$
Dream lilac $1:2.17$ (to 2 d.p.)

(b) Sugar lilac

What progress have you made? (p 16)

1 (a) 15 litres (b) 5 litres

2 (a) 20 litres (b) 6 litres

3 (a) 3 litres black, 9 litres white

4 (a) $4:3$ (b) $1:4$ (c) $2:5$ (d) $2:3$

5 (a) Harry £8, Will £4

(b) $\frac{1}{3}$

6 (a) Gavin £16, Susan £20

(b) Lee £10, Jo £30, Vi £40

7 A, with the pupil's explanation

Practice booklet

Section A (p 5)

1 (a) (i) 15 g (ii) 36 g

(b) (i) 10 g (ii) 14 g

2

Pure gold (g)	Other metals (g)
6	**10**
12	**20**
15	**25**
9	15
18	30
45	75

3

Pure gold (g)	Other metals (g)
7	**5**
21	**15**
28	20
35	**25**
49	35
126	90

Section B (p 6)

1 (a) 1 to 2 (b) 3 to 1 (c) 1 to 4
(d) 3 to 2 (e) 4 to 3

2 (a) C
   (b) (i)   D, 3 to 2
        (ii)   A, 1 to 3
        (iii) C, 2 to 1
        (iv)   B, 4 to 3
        (v)   D, 3 to 2

Section E (p 7)

1 (a) 4:3   (b) 3:4

2 (a) 2:3   (b) 3:1   (c) 3:5
   (d) 5:2   (e) 7:6

3 (a) 2:3   **6**:9   10:**15**
   (b) 4:1   **12**:3   8:**2**
   (c) 6:2   **3**:1   18:**6**
   (d) 4:10   2:**5**   **6**:15
   (e) 30:40 **6**:8   3:**4**
   (f) 7:3   21:**9**   **28**:12

4 20:30 and 6:9
   21:6 and 7:2
   9:3 and 12:4
   3:5 and 12:20
   12:9 and 8:6
   8:20 and 2:5

5 (a) 9   (b) 4
   (c) 15 loam and 5 sand

6 (a) 4 litres   (b) 12 litres
   (c) 15 blue and 10 red

Section F (p 8)

1 (a) 24 and 12   (b) $\frac{1}{3}$

2 (a) Jenny £10, Helen £40
   (b) $\frac{1}{5}$ (c) $\frac{4}{5}$

3 (a) £4 and £6   (b) £10 and £15
   (c) £32 and £48   (d) £17 and £25.50

4 (a) £10   (b) £12
   (c) Paul £18.75, Anna £11.25

5 15

6 Claire 14 sweets, Emma 7 sweets

7 Gert 4 sweets, Hans 12 sweets,
   Ivy 8 sweets

8 (a) £4, £4, £12   (b) £10, £5, £20
   (c) £20, £30, £50   (d) £9, £36, £54

Section G (p 9)

1 (a) 2:3 = 4:**6**   (b) 1:5 = **5**:25
   (c) 4:7 = 12:**21**

2 (a) Pastel pink   (b) Gun grey
   (c) Sky blue

3 (a) Lime Fizz   1:2.4
       Tangy Lime   1:2.25
   (b) Tangy Lime

4 Flaky pastry

③ Starting equations

	Optional
Practice booklet pages 10 to 12	OHP transparency of sheet 211

Ⓐ **Review** (p 17)

This section reviews the concrete idea of balancing and solving equations in the form $w + w + w = w + 20$.

> Optional: OHP transparency of sheet 211, cut into separate pieces.

'Used multiple layered OHTs to show lifting off. Went really well.'

◊ Ask pupils to find the weight of a shell and the weight of a can. In discussion, emphasise what you can do to both sides and still leave the scales in balance.

◊ If pupils do not suggest it themselves, remind them how to write an equation for each picture. Using w for the weight of a shell in grams, the equation is $w + w + w = w + 20$. (There is no need to discuss the shorthand form $3w = w + 20$ yet as this is covered in section B. However, some pupils may be ready to use it.) Encourage them to write the intermediate steps as equations too, in this case giving $w + w = 20$ and finally $w = 10$.

Stress that the letter chosen stands for the *weight of the object in grams*. It is not an abbreviation for the name of the object.

◊ Sheet 211 has three 5 gram weights so that the 10 in picture 2 can be split into two 5s if wished.

A1 Encourage pupils to begin each problem by stating what their chosen letter stands for.

B Shorthand (p 18)

The emphasis here is on using shorthand and setting out solutions clearly.

◊ Arrows and 'bubbles' are used here to show what happens to each side of the equation each time and pupils can show what they do to both sides in a similar way. Another way to set out solutions is to write what is done to both sides in brackets at the side each time and this way of setting out will be used later in the course.

Emphasise that it is best to

- write only one equals sign per line
- keep equals signs underneath each other
- always do a check.

◊ In the frogs example and in questions B1 to B3, pupils form an equation from a puzzle. In these cases, suggest that pupils check by working from the original puzzle *and* by substituting in the equation itself. This may help to identify any mistake in writing down the original equation.

Equation bingo

'Took time to play this game which was really good.'

◊ This game provides practice in solving equations.

Suggested equations are shown below. If you write the equations on the board one at a time you can see which you have used, and pupils can return to any for which they need more time.

$5p + 1 = 2p + 4$	$(p = 1)$	$6r = 2r + 8$	$(r = 2)$
$2q + 6 = 4q$	$(q = 3)$	$2w + 5 = w + 10$	$(w = 5)$
$4a + 1 = 2a + 13$	$(a = 6)$	$q + 31 = 5q + 3$	$(q = 7)$
$10x + 36 = 136$	$(x = 10)$	$3s = s + 26$	$(s = 13)$
$4t = 3t + 15$	$(t = 15)$	$s + 65 = 4s + 5$	$(s = 20)$

C Making equations (p 20)

◊ Pupils may find it helps to copy the expressions onto pieces of paper.

◊ You may need to point out that reversing the sides of an equation does not produce a different equation. For example, the equation $19 + 4x = 3x + 25$ is essentially the same as $3x + 25 = 19 + 4x$.

D Thinking of a number (p 20)

In this section, pupils form equations for themselves. It may be found difficult by weaker pupils.

◊ You could discuss the Adie and Jaspaal problem with the whole class and then split into groups. One person from each group might then tell the whole class how their group solved the problem. By the end of the discussion, pupils should see that forming an equation and solving it is one way to solve this type of puzzle.

◊ Again, suggest that pupils check by working from the original puzzle *and* by substituting in the equation itself.

Ⓐ Review (p 17)

Here x has been used as the unknown but pupils can use any letter each time.

A1 (a) $x + x + x + 4 = 10 + x$
$x = 3$ and the pupil's check

(b) $x + x + x + x + x + 2 = x + x + 20$
$x = 6$ and the pupil's check

(c) $x+x+x+x+x+x+5=x+x+x+x+17$
$x = 6$ and the pupil's check

(d) $x + x + 15 = x + x + x + x + 8$
$x = 3.5$ and the pupil's check

(e) $x + 27 = x + x + x + x + 3$
$x = 8$ and the pupil's check

(f) $x+x+x+x+x+x+x+4=x+x+24$
$x = 4$ and the pupil's check

Ⓑ Shorthand (p 18)

B1 (a) $5p + 2 = 2p + 20$

(b) $p = 6$

(c) $5 \times 6 + 2 = 32$
$2 \times 6 + 20 = 32$ so both sides agree

B2 $3x + 4 = 6x + 1$
$x = 1$
Check:
$3 \times 1 + 4 = 7$
$6 \times 1 + 1 = 7$ so both sides agree

B3 $4x + 90 = 2x + 400$
$\qquad x = 155$
Check:
$4 \times 155 + 90 = 710$
$2 \times 155 + 400 = 710$ so both sides agree

B4 In each of these solutions, unknowns have been taken off each side first.
A constant can be taken off first.

(a) $4c + 1 = 3c + 8$
$\qquad c + 1 = 8$
$\qquad\quad c = 7$
Check:
$4 \times 7 + 1 = 29$
$3 \times 7 + 8 = 29$ so both sides agree

(b) $3d + 28 = 5d + 8$
$\qquad\quad 28 = 2d + 8$
$\qquad\quad 20 = 2d$
$\qquad\qquad d = 10$
Check:
$3 \times 10 + 28 = 58$
$5 \times 10 + 8 = 58$ so both sides agree

(c) $37 + 2n = 17 + 4n$
$\qquad 37 = 17 + 2n$
$\qquad 20 = 2n$
$\qquad\quad n = 10$
Check:
$37 + 2 \times 10 = 57$
$17 + 4 \times 10 = 57$ so both sides agree

(d) $5t + 7 = 27$
$\qquad 5t = 20$
$\qquad\quad t = 4$
Check:
$5 \times 4 + 7 = 27$ so both sides agree

(e) $5y = 4y + 5$
$\qquad y = 5$
Check:
$5 \times 5 = 25$
$4 \times 5 + 5 = 25$ so both sides agree

(f) $4x + 20 = 8x$
$20 = 4x$
$x = 5$
Check:
$4 \times 5 + 20 = 40$
$8 \times 5 = 40$ so both sides agree

(g) $5q + 11 = 3q + 21$
$2q + 11 = 21$
$2q = 10$
$q = 5$
Check:
$5 \times 5 + 11 = 36$
$3 \times 5 + 21 = 36$ so both sides agree

(h) $6m + 2 = 14 + 2m$
$4m + 2 = 14$
$4m = 12$
$m = 3$
Check:
$6 \times 3 + 2 = 20$
$14 + 2 \times 3 = 20$ so both sides agree

B5 (a) $7 + 2n = 22$
$2n = 15$
$n = 7.5$
Check:
$7 + 2 \times 7.5 = 22$ so both sides agree

(b) $21 + 3m = 40 + m$
$21 + 2m = 40$
$2m = 19$
$m = 9.5$
Check:
$21 + 3 \times 9.5 = 49.5$
$40 + 9.5 = 49.5$ so both sides agree

(c) $2f + 37 = 1 + 12f$
$37 = 1 + 10f$
$36 = 10f$
$f = 3.6$
Check:
$2 \times 3.6 + 37 = 44.2$
$1 + 12 \times 3.6 = 44.2$ so both sides agree

(d) $3h + 10 = 8h + 1$
$10 = 5h + 1$
$9 = 5h$
$h = 1.8$

Check:
$3 \times 1.8 + 10 = 15.4$
$8 \times 1.8 + 1 = 15.4$ so both sides agree

(e) $5p + 4 = p + 41$
$4p + 4 = 41$
$4p = 37$
$p = 9.25$
Check:
$5 \times 9.25 + 4 = 50.25$
$9.25 + 41 = 50.25$ so both sides agree

(f) $3r + 31 = 5 + 8r$
$31 = 5 + 5r$
$26 = 5r$
$r = 5.2$
Check:
$3 \times 5.2 + 31 = 46.6$
$5 + 8 \times 5.2 = 46.6$ so both sides agree

B6 The pupil's equation and solution

ℂ **Making equations** (p 20)

C1 (a) $x = 11$

(b) $x = 6$

(c) The pupil's equation with its solution

(d) The pupil's two equations with their solutions

(The remaining four equations and their solutions are

$3x + 25 = 31 + x$ $\qquad x = 3$

$5x + 3 = 19 + 4x$ $\qquad x = 16$

$5x + 3 = 31 + x$ $\qquad x = 7$

$19 + 4x = 31 + x$ $\qquad x = 4$)

C2 The six possible equations and their solutions are

$7n + 9 = 11 + 6n$ $\qquad n = 2$

$7n + 9 = 41 + 3n$ $\qquad n = 8$

$7n + 9 = 5n + 17$ $\qquad n = 4$

$11 + 6n = 41 + 3n$ $\qquad n = 10$

$11 + 6n = 5n + 17$ $\qquad n = 6$

$41 + 3n = 5n + 17$ $\qquad n = 12$

D Thinking of a number (p 20)

D1 (a) $4n = 2n + 12$

(b) $n = 6$ and the pupil's check

D2 (a) $10n = 5n + 35$

(b) $n = 7$

D3 (a) $4n + 9 = 6n$

(b) $n = 4.5$

D4 $6n + 8 = 2n + 40$
$n = 8$

*D5 (a) $6(n + 3) = 4(n + 8)$

(b) $6n + 18 = 4n + 32$
$n = 7$

*D6 $4(n + 3) = 5n + 3$
$4n + 12 = 5n + 3$
$n = 9$

*D7 $7n + 0.6 = 5(n + 1.8)$
$7n + 0.6 = 5n + 9$
$n = 4.2$

*D8 The pupil's problems

What progress have you made? (p 22)

1 (a) $t = 11$ (b) $g = 12$ (c) $r = 8$

2 $7n = 5n + 18$
$n = 9$

Check:
$7 \times 9 = 63$
$5 \times 9 + 18 = 63$ so both sides agree

Practice booklet

Section A (p 10)

1 (a) $x + x + x + x + x + 8 = 68$
$x = 12$ and the pupil's check

(b) $x + x + x + x + x + 2 = x + x + 17$
$x = 5$ and the pupil's check

(c) $x + x + x + x + 31 =$
$x + x + x + x + x + x + 7$
$x = 12$ and the pupil's check

(d) $x + x + x + x + x + 27 =$
$x + x + x + x + x + x + x + x + x + 7$
$x = 5$ and the pupil's check

(e) $x + x + x + 18 = x + x + x + x + x + 3$
$x = 7.5$ and the pupil's check

Section B (p 11)

Each solution should include the pupil's check.

1 (a) $n = 2$ (b) $w = 6$ (c) $h = 1$

(d) $t = 4$ (e) $x = 4$ (f) $p = 4$

(g) $d = 8$ (h) $m = 3$ (i) $k = 7$

(j) $m = 2$ (k) $p = 4$ (l) $y = 3$

2 (a) $t = 9$ (b) $t = 2$ (c) $t = 3$

(d) $t = 3$ (e) $t = 5.5$ (f) $t = 1.5$

(g) $t = 10$ (h) $t = 1.2$ (i) $t = 0$

Section C (p 11)

1 $x = 2$

2 $x = 5$

3 The remaining four equations are
$2x + 11 = x + 19$ $x = 8$
$2x + 11 = 3x + 9$ $x = 2$
$x + 19 = 7x + 1$ $x = 3$
$3x + 9 = 7x + 1$ $x = 2$

Section D (p 12)

1 $10n + 3 = 6n + 27$ $n = 6$

2 $4n + 80 = 20n + 48$ $n = 2$

3 $6n + 6 = 2n + 16$ $n = 2.5$

*4 $3(n + 5) = 2(n + 6)$
$3n + 15 = 2n + 12$ $n = {}^-3$

④ Cuboids

As well as introducing volume, this gives plenty of practice in mental multiplication, since only a few of the questions need a calculator.

T	p 23	**A** How many cubes?
	p 24	**B** Volume of a cuboid
	p 26	**C** Shapes made from cuboids
T	p 28	**D** Litres and millilitres
T	p 29	**E** Cubic metres

> **Optional**
> Triangular dotty paper
> Multilink cubes
>
> **Practice booklet** pages 13 to 15 (needs triangular dotty paper)

Ⓐ How many cubes? (p 23)

> Optional: triangular dotty paper, multilink cubes

This introduces the concept of volume and reinforces earlier work on factors.

◊ You can ask the class to suggest everyday objects that are roughly cuboids. Establish that a cube is a special kind of cuboid.

◊ Some pupils may be helped by having multilink cubes for this section.

Ⓑ Volume of a cuboid (p 24)

The aim is that pupils move naturally away from counting cubes and thinking about layers of cubes to seeing that there is a general rule for the volume of a cuboid.

◊ The questions from B4 onwards can help pupils make sense of decimal multiplication on a calculator.

◊ You can also refer to some of the diagrams in this section to revise area (the areas of separate faces and the total surface area of a cuboid), in preparation for the surface area work in section C.

If pupils feel confident about finding volumes from decimal lengths, they could use a calculator or spreadsheet to explore these questions.

- What edge length must a cube have if its volume is 100 cm³?
- A cube is 5 cm by 5 cm by 5 cm. What edge length is needed for a cube with twice the volume of this one?

ℂ **Shapes made from cuboids** (p 26)

◊ Pupils have to be systematic to find the total surface area, as C2 demonstrates. Another approach is to sketch a net of the solid, labelling its edge lengths and on each face recording its area.

𝔻 **Litres and millilitres** (p 28)

◊ Pupils could measure edge lengths of 1 litre cuboid juice containers and calculate their capacity. The results are likely to be just over 1 litre and you can ask them to explain why.

𝔼 **Cubic metres** (p 29)

Pupils can make a cube from rolled up newspaper like the one in the photograph. This can help them visualise how many such cubes would fit into their classroom, the container on a lorry, a home freezer or some other such space. World records for the number of people fitted into a given space usually appeal to pupils and could generate some mathematical discussion.

𝔸 **How many cubes?** (p 23)

A1 (a) 16   (b) 28   (c) 27
     (d) 72   (e) 80

A2 (a) 60   (b) 175   (c) 5760

A3 12 by 1 by 1, 4 by 3 by 1, 3 by 2 by 2
(The three numbers can of course be in any order.)

A4 (a) 8 by 1 by 1, 4 by 2 by 1, 2 by 2 by 2
     (b) 20 by 1 by 1, 10 by 2 by 1,
         5 by 4 by 1, 5 by 2 by 2

A5 Only one, 23 by 1 by 1, since 23 is a prime number.
Numbers with more factors give more cuboids.

𝔹 **Volume of a cuboid** (p 24)

B1 (a) 40 cm³   (b) 48 cm³   (c) 54 cm³

B2 (a) 30 cm³   (b) 16 cm³   (c) 60 cm³
     (d) 48 cm³   (e) 24 cm³   (f) 125 cm³

B3 (a) 60 cm³   (b) 36 cm³   (c) 28 cm³
     (d) 36 cm³

B4 (a) volume = length × width × height
     (b) The volume is 10.5 cm³. This agrees with the diagram, where there are 9 full centimetre cubes and 3 half centimetre cubes.

B5 (a) 36 cm³   (b) 45 cm³

B6 12.25 cm³

B7 (a) $70 \, cm^3$ (b) $20.25 \, cm^3$

B8 $a = 4$ cm, $b = 8$ cm, $c = 1.5$ cm, $d = 4$ cm

B9 0.5 cm

© **Shapes made from cuboids** (p 26)

C1 (a) $24 \, cm^3$ (b) $60 \, cm^3$ (c) $84 \, cm^3$

C2 (a) (i) $42 \, cm^2$ (ii) $42 \, cm^2$
         (iii) $8 \, cm^2$ (iv) $6 \, cm^2$
         (v) $4 \, cm^2$ (vi) $10 \, cm^2$
         (vii) $12 \, cm^2$ (viii) $16 \, cm^2$
      (b) $140 \, cm^2$

C3 (a) $6 \, cm^3$ (b) $24 \, cm^2$

C4 (a) $28 \, cm^3$ (b) $58 \, cm^2$

Ⓓ **Litres and millilitres** (p 28)

D1 4

D2 10

D3 (a) 2 litres   (b) 3 litres   (c) 9 litres
      (d) 20 litres   (e) 672 litres

D4 (a) 2 litres   (b) 3.5 litres   (c) 0.5 litre
      (d) 0.85 litre

D5 (a) 20   (b) 150   (c) 200   (d) 100

D6 (a) 400 ml   (b) 650 ml   (c) 273 ml
      (d) 50 ml

D7 60 ml, 0.5 litre, 750 ml, $1100 \, cm^3$, 1.2 litres

D8 (a) 0.3 litre   (b) 0.45 litre   (c) 0.05 litre
      (d) 0.001 litre

Ⓔ **Cubic metres** (p 29)

The iceberg dimensions give a volume of $1347.5 \, km^3$, but as this is an estimate, the volume can be given as $1300 \, km^3$.

What progress have you made? (p 29)

1 (a) $1600 \, cm^3$ (b) $30 \, m^3$

2 (a) $10 \, cm^3$ (b) $30 \, cm^2$

3 $70 \, cm^3$, 0.1 litre, 105 ml, $550 \, cm^3$, 0.65 litre, 1200 ml, 1.7 litres

Practice booklet

Section A (p 13)

1 The pupil's sketch, which could be of any one of these cuboids:
   30 by 1 by 1, 15 by 2 by 1, 10 by 3 by 1, 5 by 3 by 2

2 The pupil's sketches of these cuboids:
   32 by 1 by 1, 16 by 2 by 1, 8 by 4 by 1, 8 by 2 by 2, 4 by 4 by 2

3 (a) 40   (b) 60   (c) 36   (d) 96

Section B (p 14)

1 (a) $12 \, cm^3$ (b) $120 \, cm^3$ (c) $24 \, cm^3$

2 (a) $15 \, cm^3$ (b) $20 \, cm^3$ (c) $108 \, cm^3$
   (d) $18 \, cm^3$ (e) $70 \, cm^3$

3 (a) $208 \, cm^3$ (b) $675 \, cm^3$ (c) $1700 \, cm^3$
   (d) $189 \, cm^3$

4 $a = 2$ cm, $b = 2.5$ cm, $c = 5$ cm

Section C (p 15)

1 (a) $200 \, cm^3$ (b) $220 \, cm^2$

Section D (p 15)

1 120

2 (a) 0.57 litre   (b) 0.35 litre
   (c) 0.007 litre   (d) 0.032 litre

3 48 ml, 305 ml, 0.4 litre, $410 \, cm^3$, 3.5 litres

4 (a) 200 ml   (b) 700 ml   (c) 750 ml
   (d) 400 ml

Graphs, charts and tables

Pupils interpret different types of graphs and charts from real life
sources and draw their own.

Essential	Optional
Graph paper, squared paper	*Wall's monitor* (available from Birds Eye Wall's Limited, Station Avenue, Walton-on-Thames, Surrey, KT12 1NT; tel. 01932 263000; this was free at the time of writing)

Practice booklet pages 16 to 20

Ⓐ **Regional differences** (p 30)

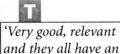

'Very good, relevant and they all have an opinion.'

'Wall's info led to considerable useful discussion and caught the pupils' interest.'

◊ The chart shows the total weekly income, which is the sum of pocket money, other handouts from relatives and any earnings. You may want to collect data from the class to compare with the data given in the charts. Income can be a sensitive issue so to avoid a show of hands you could get pupils to write their income anonymously on a piece of paper which you collect.

◊ The data from *Wall's monitor* could be used when drawing graphs and charts later in the unit. It is an annual survey based on data provided by Gallup. It contains a host of data on children's income.

Ⓑ **Changes over time** (p 31)

Graph paper

'This worked well as a group activity'.

◊ There are many other questions you could ask about the graph of sales of recorded music. Some pupils may not know what an LP or a single is.

◊ Questions B15 to B17 discuss the significance of a line joining data points. In questions B11 and B14 it would be reasonable to join points with a smooth curve.

ℂ **Frequency charts and tables** (p 34)

> Squared paper

This section provides data which you can use for class lessons on the techniques of drawing graphs and charts. Be prepared to spend some time on this work: classifying the data and drawing the graphs and charts takes quite a while.

If you use an OHP, you will find it useful to have an acetate sheet ruled like the graph paper the pupils use. Otherwise, a grid on the board is helpful.

◊ The steps involved in drawing a bar chart are summarised in the yellow box. You can try out different widths of class interval with pupils so they can see the effects.

◊ For discrete data (as in the case of the pupils' test scores data) class intervals can be marked on the axis with gaps.

For continuous data (such as the race times), a continuous scale should be used.

The problem about values that are used on the boundary does not arise in the race times data set. This issue can be left to unit 26, 'Distributions'.

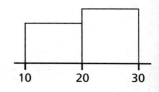

C4 Check whether pupils are reading the scales correctly (the vertical scale is marked every 2%). Pupils' own data on time spent watching TV could be used for comparison.

𝔸 **Regional differences** (p 30)

A1 (a) 605p

(b) Scotland

(c) North West

(d) 605p − 397p = 208p

A2 (a) Welsh

(b) Thames, Southern, South Western and Welsh

(c) Northumbria & Yorks, South Western, Welsh

(d) Severn

(e) Anglian

(f) Anglian because it has the smallest percentage of 'good' samples

or

North West because it has the largest percentage of 'bad' samples

𝔹 **Changes over time** (p 31)

B1 (a) 1990

(b) (i) 1987 and 1988 (ii) About 60 000

(c) 1985 and 1986 (d) 1989 and 1990

B2 12 000 (roughly)

B3 For example, 'Between 1983 and 1985 complaints rose steadily. Then they dropped between 1985 and 1986. They rose in the next year and stayed the same between 1987 and 1988. They rose again between 1988 and 1989 and then rose very steeply up to 1990.'

B4 The scale is not continuous from zero.

B5 Between 1985 and 1986

B6 1986

B7 (a) The sales of CDs started in 1983 and went up every year.
(b) The sales of cassettes rose to just over 80 million in 1989, then dropped until 1992, then stayed roughly the same to 1995.
(c) In 1989   (d) 1986
(e) 80 million

B8 (a) Singles   (b) Cassettes (c) CDs

B9 About 160 million

B10 1998

B11 (a)

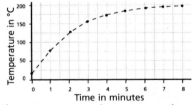

(b) The temperature increases quite quickly at first. Later it continues to increase but the rate of increase gradually becomes less.

B12 (a)

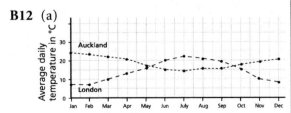

(b) The Auckland temperatures are generally higher than the London temperatures. London is warmest in the middle of the year. Auckland is coolest then.

(c) London   (d) Auckland

B13

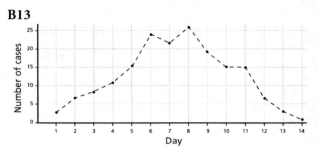

B14

A graph like this could be joined with a smooth curve.

B15 No, you would expect it to be much cooler at midnight. The dotted lines merely help you 'take in' the graph.

B16 Yes

B17 Yes

ℂ **Frequency charts and tables** (p 34)

Each bar chart depends on the class intervals chosen. The data has been put in order here, however, to help you check pupils' work.

C1 2, 3, 4, 9
11, 13, 14, 16, 18, 19
22, 22, 23, 25, 27
33, 34, 37, 38, 38, 38
40, 41, 41, 44, 44, 45, 46,
50, 51, 52, 53, 56, 59
60, 61, 64, 64, 66, 69
70, 71, 72, 72, 75, 77
80, 88

C2 1.64 1.65 1.81 1.94 1.95
2.05 2.06 2.27 2.40 2.45
2.55 2.56 2.59 2.69 2.71 2.88 2.90 2.96
3.11 3.19 3.27 3.37 3.47 3.47
3.50 3.70 3.71 3.77
4.08 4.39

C3 The pupil's table and answer

C4 (a) More boys (b) More boys
    (c) 15% (d) Up to 1 hour
    (e) Year 10 (from the general shape of the graphs)

What progress have you made? (p 36)

1

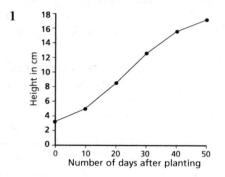

    (a) 10.6 cm (b) About 18 cm

2 (a) September

    (b) Between April and May

3 This is a possible answer.

Mark	Frequency
10–14	3
15–19	7
20–24	7
25–29	6
30–34	6
35–39	6
40–44	4
45–49	1

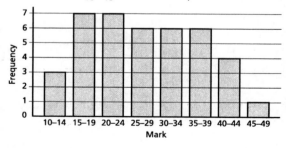

4 The girls are generally taller than the boys (from the shapes of the graphs).

Practice booklet

Section A (p 16)

1 (a) The UK (b) Greece
    (c) Portugal and Greece
    (d) The Netherlands
    (e) Germany
    (f) Spain, Portugal, Italy and Germany
    (g) Spain, Portugal, Italy, Ireland, Greece and Germany.

2 (a) Bus (b) Car (c) 30%
    (d) 8% (e) Car (f) Bike
    (g) More than year 7
    (h) More than year 7

Section B (p 18)

1 (a) London, 4°C (b) Sydney, 22°C
    (c) The temperature went down until the 7th and 8th, then went up until the 11th, and then went down again.
    (d) On the 4th, 5th, 6th, 7th, 8th, 13th, 14th and 15th
    (e) On the 1st, difference 8 degrees
    (f) 10°C (g) On 6 days

Section C (p 19)

1 (a) Year 7 (b) 31 to 60 (c) 5%
    (d) 10% (e) 16–17% (f) 42–43%
    (g) Year 11; they have more high bars to the right.

2 The pupil's graphs

This is the most obvious grouping for part (a).

Weight	Frequency
0–9	2
10–19	8
20–29	6
30–39	12
40–49	14
50–59	8

Fractions, decimals and percentages

Throughout the unit, percentages are expected to be given to the nearest whole number. You can ask pupils to give more accurate answers where you think this is appropriate.

If no year is stated, the information given relates to the late 1990s.

p 37 **A** Review	Finding a fraction of an amount Expressing one number as a fraction of another Converting between fractions, percentages and decimals Finding a percentage of a quantity Writing one number as a percentage of another
p 39 **B** Mixed percentage problems	
p 40 **C** A load of rubbish	Using percentages to analyse data in the context of waste and recycling

---

Essential

Compasses and angle measurers
or pie chart scales (for drawing pie charts in C6)

Practice booklet pages 21 to 24

Ⓐ **Review** (p 37)

Ⓑ **Mixed percentage problems** (p 39)

B5 The figures in the table are taken from the *Digest of Environmental Statistics* no. 20, 1998, the Department of the Environment, Transport and the Region's annual reference volume for environmental data. The Digest is available as a book published by The Stationery Office.

Ⓒ **A load of rubbish** (p 40)

A pie chart scale or compasses and angle measurer for C6

C6 Pupils can use a pie chart scale and work with percentages to draw their pie charts. Alternatively, they can calculate angles and use a protractor or angle measurer.

For Japan, rounding the percentages to the nearest 1% leads to a total of 99% (40% + 18% + 40% + 1%). Pupils may already have experience of totals of, say, 99% or 101% when analysing data. In this case, the figures are rounded down each time leading to a total of 99%. Pupils may need to consider how to deal with this when drawing the pie chart.

A possible strategy is to try to work with percentages correct to one decimal place (40.4%, 18.3%, 40.4%, 0.9%).

In calculating the percentage of Japan's waste that is composted, pupils may obtain a decimal result in standard form on their calculators. They could try multiplying by 100 to convert it to a percentage. Alternatively, they could 'fix' the number of decimal places to 2 (or 3).

Ⓐ Review (p 37)

A1 (a) 3   (b) 30   (c) 21   (d) 45
    (e) 24   (f) 84   (g) 20   (h) 10

A2 (a) 2000 kg   (b) $\frac{1}{4}$
    (c) (i) True   (ii) True
       (iii) False   (iv) True
    (d) $\frac{8}{15}$

A3 (a) 0.75   (b) 0.6   (c) 0.02
    (d) 0.55   (e) 0.57

A4 (a) $\frac{1}{2}$ (b) $\frac{1}{4}$ (c) $\frac{1}{10}$
    (d) $\frac{2}{5}$ (e) $\frac{1}{20}$

A5 (a) 0.75   (b) 0.94   (c) 0.3
    (d) 0.03   (e) 0.84

A6 (a) 21%   (b) 34%   (c) 90%
    (d) 8%   (e) 20%

A7 (a) 35%   (b) 75%   (c) 4%
    (d) 40%   (e) 42%   (f) 92%
    (g) 58%   (h) 3%   (i) 44%
    (j) 73%

A8 (a) 7.7 kg (rounded from 7.68)
    (b) 4.8 tonnes (rounded from 4.75)
    (c) 279.5 g
    (d) 0.1 kg (rounded from 0.138)
    (e) 7.8 kg   (f) 5.4 kg

A9 12 g

A10 12.5 kg

A11

Bird	Number	Percentage
Sparrow	9	**21%**
Blue tit	7	**17%**
Starling	6	14%
Blackbird	5	**12%**
Chaffinch	8	**19%**
Greenfinch	7	**17%**

Ⓑ Mixed percentage problems (p 39)

B1 £16.80

B2 26%

B3 (a) £0.81 or 81p (rounded from 0.807)
    (b) £0.19 or 19p (rounded from 0.1883)
    (c) £0.19 or 19p as part (b)
    (d) £1.21 (rounded from 1.2105)

B4 35%

B5 (a) Ozone depletion
    (b) 35%   (c) 30%

Ⓒ A load of rubbish (p 40)

C1 (a) True   (b) False
    (c) False   (d) False

C2 (a) 2%

(b)

Year	% of cans recycled
1989	2%
1990	5%
1991	11%
1992	16%
1993	21%
1994	24%
1995	28%
1996	31%

C3 (a) 8%

(b)

Year	% of cans recycled
1989	8%
1990	9%
1991	10%
1992	12%
1993	13%
1994	14%
1995	14%
1996	12%

C4 The pupil's comments such as: 'The percentage of aluminium cans recycled has risen each year and over eight years has risen from 2% to 31%. The percentage of steel cans recycled rose between 1989 and 1994 but only by 1% or 2% each year. Between 1995 and 1996 the percentage fell by 1%. Over eight years the percentage has risen but only from 8% to 12%.'

C5 (a)

Waste	Average weight (kg)
Kitchen waste	108
Newspaper	70
Paper	64
Ash and dust	44
Clear glass	32
Magazines	32
Nappies, plastic film, clothing	76
Miscellaneous waste	210

(b) 4 800 000 kg or 4800 tonnes

*(c) About 7300 trees

C6 (a) 20%

(b) **USA**

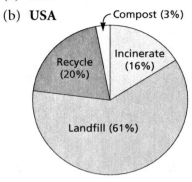

Japan

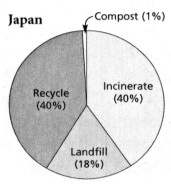

(c) The pupil's comments, for example: 'Japan recycles 40% of its waste but the USA only recycles 20%. The USA disposes of 61% of its waste in landfill sites but Japan only disposes of 18% of its waste this way.'

What progress have you made? (p 42)

1 (a) 160 kg (b) $\frac{1}{5}$

2 26 tonnes

3 Cultivated: 191 000 km^2
Grazing: 109 000 km^2
Forest: 147 000 km^2
Other: 98 000 km^2

4 Cultivated: 48%
Grazing: 18%
Forest: 22%
Other: 12%

Practice booklet

Sections A and B (p 21)

1 (a) 10 (b) 8 (c) 60
   (d) 12 (e) 36 (f) 25

2 (a) 48 (b) $\frac{1}{4}$ (c) $\frac{1}{6}$

3 (a) 135 (b) 180 (c) $\frac{1}{4}$
   (d) 1020 (e) $\frac{3}{4}$

4 (a) 0.8 (b) 0.25 (c) 0.9
   (d) 0.63 (e) 0.83 (f) 0.67

5 (a) 0.65 (b) 0.4 (c) 0.07
   (d) 0.75 (e) 0.08 (f) 0.12
   (g) 0.15 (h) 0.91 (i) 0.06

6 (a) 71% (b) 80% (c) 9%

7 (a) 75% (b) 80% (c) 70%
   (d) 15% (e) 48% (f) 55%

8 (a) 33% (b) 17% (c) 46%
   (d) 18% (e) 18% (f) 10%

9 (a) 14 kg (b) £121.80
   (c) 68.82 kg (d) 5.28 g
   (e) £12.24 (f) 354.42 cm

10 (a) £6.82 (b) 4.94 m
    (c) 43.17 kg (d) 3.66 g

11 41%

12 (a) A True B False
        C True D True
    (b) 21%
    (c) 3%

13 (a) 4% (b) £249 120
    (c) £2 117 520

Section C (p 24)

1 (a) England 9 032 000
        Wales 543 000
        Scotland 968 000
        N. Ireland 395 000
   (b) England 7 606 000
        Wales 457 000
        Scotland 764 000
        N. Ireland 189 000

2 (a) 23%
   (b)

1930	51%	12%	**25%**	12%
1950	45%	**16%**	**23%**	**17%**
1970	39%	**5%**	**10%**	**46%**
1990	**32%**	**5%**	**13%**	**50%**

   (c) The pupil's pie charts, one for each year
   (d) The pupil's comments on the changes over the years

7 Using rules

Practice booklet pages 25 and 26

A Katy's Catering (p 43)

◊ You could discuss with pupils how many of each item would be needed for a party of (say) 10 people, then 20 people etc.

B Shorthand (p 45)

This section introduces letters standing for numbers. It is very important to emphasise again that, for example, h stands for the *number of hats* and not for the word 'hats'. It helps sometimes to choose a letter that differs from the first letter of the object referred to.

◊ Discuss with pupils how to write the word-formulas in the usual mathematical shorthand. You can use the formulas on pages 43 and 44 as a basis for this, or ask pupils to suggest formulas of their own to write in shorthand.

Include discussion of substituting into the shorthand formulas, so that pupils can see how they work.

◊ Discuss common mistakes such as
  • writing $n \times 2$ as $n2$ and not (as convention demands) $2n$
  • substituting (say) $n = 3$ into $2n$ and getting 23.
  • Not realising that n and $1n$ are the same.

B7 Watch out for pupils who choose $3s = n$, simply translating the words into symbols.

C Working out formulas (p 48)

◊ To remind pupils about brackets, you may wish to discuss each of the Expocamp rules using (say) $n = 20$. (Use a multiple of 4 so that the tent rule does not give rise to difficulties.)

◊ Much of this unit lends itself to treatment on a spreadsheet. For example:

	A	B	
1	Katy's Catering		
2	Number of people	10	
3			
4	Chicken legs	=B2+6	
5	Paper cups	=B2*2	
6	Lemonade	=B2/2	
7	Sausage rolls	=3*B2+10	

Getting the pupils themselves thinking about what formulas are appropriate can help their understanding of the algebra.

Ⓐ Katy's Catering (p 43)

A1 (a) (i) 28   (ii) 70   (iii) 160
    (b) 30 people

A2 (a) (i) 16   (ii) 46   (iii) 31
    (b) 54 people

A3 (a) 320   (b) 130 people

A4 (a) (i) 5   (ii) 10
    (b) 60 people

A5 (a) (i) 14   (ii) 24
    (b) 18 people

A6 (a) 10   (b) 8   (c) 50

Ⓑ Shorthand (p 45)

B1 (a) 18   (b) 26   (c) 38

B2 (a) 24   (b) 30   (c) 60

B3 (a) 160   (b) 45   (c) 8

B4 (a) 70   (b) 55   (c) 160

B5 (a) 55   (b) 115
    (c) You need **5** iced buns for each person, and another **15** extra.

B6 $c = n + 6$

B7 (a) C   (b) $s = 3n$

B8 (a) Number of loaves of bread needed = number of people ÷ 2
    (b) $b = \frac{n}{2}$

B9 Pupils may have chosen different letters in these rules. They should state what each letter stands for.
    (a) $l = n + 2$ (b) $d = 3n - 6$
    (c) $p = 4n + 8$ (d) $f = 2n + 10$

B10 Pupils may have chosen different letters in these rules. They should state what each letter stands for.
    (a) $m = 6n$ (b) $r = 2n + 6$
    (c) $c = n + 10$ (d) $t = \frac{n}{6}$
    (e) $s = 2n + 6$

Ⓒ Working out formulas (p 48)

C1 (a) 12 tents
    (b) 156 sardine cans, 84 ropes, 482 packets of porridge

C2 (a) 5 tents, 44 sardine cans, 28 ropes, 146 packets of porridge
    (b) 27 tents, 396 sardine cans, 204 ropes, 1202 packets of porridge

C3 (a) 50 textbooks, 360 mirrors, 1000 cubes, 10 wallcharts
    (b) 25 textbooks, 160 mirrors, 600 cubes, no wallcharts
    (c) 250 textbooks, 1960 mirrors, 4200 cubes, 90 wallcharts

C4 If p is the number of pencils she buys, then $p = 2c + 50$.

C5 (a) 30 (b) 45 (c) 18 (d) 15

C6 (a) 34 (b) 4 (c) 0 (d) 194

C7 (a) 4 (b) 19 (c) $6\frac{1}{2}$ (d) 244

C8 (a) 2 (b) 10 (c) 1

C9 (a) 36 (b) 25 (c) 10

C10 (a) 19 (b) 35 (c) 11

C11 (a) 35 (b) 63 (c) 0

C12 (a) 31 (b) 9

C13 (a) 30 (b) 46 (c) 36 (d) 40
 (e) 17 (f) 7 (g) 55 (h) 4

C14 (a) 90 (b) 35 (c) 50

C15 (a) 2 (b) 2 (c) 45

What progress have you made? (p 50)

1 54

2 8

3 30

4 11

5 (a) 28 (b) 0 (c) 388

6 (a) 1 (b) 3 (c) 25

Practice booklet

Section B (p 25)

1 (a) The pupil's sentences such as:
  The number of easels is two more
  than the number of people.
  Take four brushes for each person,
  plus an extra five.
 (b) 7 easels, 25 brushes
 (c) 12 (d) 29 (e) 18

2 (a) $w = \frac{n}{4}$
 (b) 8 (c) 20 (d) 4

3 (a) $t = 6n + 4$
 (b) $s = 3n$
 (c) $p = \frac{n}{4} + 1$
 (Pupils may choose different letters. They
 should state what each letter stands for.)

Section C (p 26)

1 (a) 50 atlases, 7 globes, 45 maps
 (b) 200 atlases, 22 globes, 195 maps
 (c) 300 atlases, 32 globes, 295 maps

2 (a) 6 (b) 12 (c) 42

3 (a) 10 (b) 0 (c) 200

4 (a) 16 (b) 1 (c) 144

5 (a) 32 (b) 96 (c) 12

6 (a) 5 (b) 10 (c) 18
 (d) 1 (e) $8\frac{1}{2}$ (f) 36

7 (a) 18 (b) 3 (c) 0

Review 1 (p 51)

1 (a) John £20, Adrian £40   (b) $\frac{1}{3}$

2 (a) $c = 7$ (b) $d = 6$ (c) $e = 8$

3 10

4 (a) A: $1200\,\text{cm}^3$ B: $300\,\text{cm}^3$
    (b) $2500\,\text{ml}$
    (c) No, $2500\,\text{ml} = 2500\,\text{cm}^3$

5 (a) $1:4$ (b) $5:2$ (c) $9:8$ (d) $4:5$

6 (a) $10:10$ (b) $10:20$ or $10:21$
    (c) More were leaving.

7 $2.85\,\text{g}$

8 Yellow $50\,\text{ml}$, blue $125\,\text{ml}$, black $25\,\text{ml}$

9 $y = 56$

10 (a) and (b)

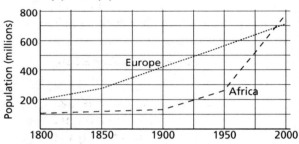

    (c) About 1990
    (d) For example, 'The population of
       Africa stayed much the same for a
       long time, then increased rapidly,
       whereas the population of Europe
       rose steadily all the time.'

11 (a) $25\,\text{kg}$ (b) $100\,\text{kg}$

12 85%

13 (a) $23:2$ (b) $6\,\text{g}$

14 $5x + 3 = 3x + 21$ $x = 9$

15 (a) $p = 50$ (b) $m = 6$

16 (a) $130\,\text{cm}^3$ (b) $202\,\text{cm}^2$

Mixed questions 1 (Practice booklet p 27)

1 £40 : £100

2 9 500 000

3 (a) 150   (b) 75%

4 $500\,\text{ml}$

5 (a) 26 kebabs   (b) 15 people

6 $8n + 21 = 11n$ $n = 7$

7 (a) $12\,000\,\text{cm}^3$ (b) $3400\,\text{cm}^2$

8 (a) $c = 16$ (b) $y = 55$ (c) $r = 47$

9 A $3:5 = 24:40$ B $5:8 = 25:40$
   So B has stronger flavour.

10 (a) $x = 9$ (b) $t = 8$ (c) $g = 6$

11 (a) $h = 3\,\text{cm}$ (b) $w = 5\,\text{cm}$

12 $8:4:3$

13 (a) 0.7 million or 700 000
    (b) 1976 or there about
    (c) 1921
    (d) The population was probably
       increasing. (We cannot be certain
       from these graphs because they do
       not give us information about
       immigration or emigration.)
    (e) The population may start to
       decrease.

Decimals and area

T

p 53	**A** Area of a rectangle	Sides to one place of decimals
p 57	**B** Areas of other shapes	Right-angled triangle and composite shapes

Practice booklet page 29

T

Ⓐ **Area of a rectangle** (p 53)

The aim is to help pupils make sense of areas from decimal lengths, rather than just substituting into formulas without thinking.

◊ The posters on page 53 can be used as the basis for teacher-led question and answer. The aim is to give pupils a feel for the fact that varying either the width or the height of the poster affects its area and hence the cost of putting it on the poster site for a week. If you record the cost of each poster on the board as you go, pupils can use these costs to work out other costs (for example, E can be worked out from A, and I can be worked out from A and E). Also you can ask whether, for example, it 'looks right' that D and E have the same cost.

The costs are A £0.80, B £0.60, C £0.90, D £0.40, E £0.40, F £0.30, G £0.45, H £0.20, I £1.20, J £0.90, K £1.35, L £0.60

◊ Although the work in A3 to A17 is about calculating area it is also designed to reveal misconceptions about decimals and to reinforce understanding of them. Pupils need to understand confidently that

$$0.46 = 4 \text{ tenths} + 6 \text{ hundredths} = 46 \text{ hundredths}$$

and that just because $0.4 \times 0.6 = 0.24$, it does not follow that $0.2 \times 0.3 = 0.6$.

Ⓑ **Areas of other shapes** (p 57)

◊ Pupils will have done work on this in year 7, but with whole-number lengths. Stress the need to sketch each composite shape, divide it into the rectangles or triangles needed and record the dimensions that need to be multiplied.

Ⓐ Area of a rectangle (p 53)

A1 £0.35

A2 (a) £3.00   (b) £0.75   (c) £1.40

    (d) £0.15

A3 (a) $0.7\,m^2$ (b) $0.4\,m^2$ (c) $0.6\,m^2$

A4 (a) Seven of the small squares

    (b) Three of the small squares

A5 P: $\frac{4}{100}\,m^2$, $0.04\,m^2$

    Q: $\frac{40}{100}\,m^2$, $0.4\,m^2$

    R: $\frac{16}{100}\,m^2$, $0.16\,m^2$

A6 (a) (i) $\frac{6}{100}\,m^2$ (ii) $0.06\,m^2$

    (b) (i) $\frac{8}{100}\,m^2$ (ii) $0.08\,m^2$

    (c) (i) $\frac{10}{100}\,m^2$ (ii) $0.1\,m^2$

A7 $\frac{37}{100}\,m^2$

A8 (a) (i) $0.85\,m^2$ (ii) $\frac{85}{100}\,m^2$

    (b) (i) $0.44\,m^2$ (ii) $\frac{44}{100}\,m^2$

    (c) (i) $0.43\,m^2$ (ii) $\frac{43}{100}\,m^2$

A9 (a) 28 hundredths

    (b) 0.28

    (c) The calculator result should be the same, 0.28

A10 (a) (i) 48 hundredths   (ii) $0.48\,m^2$

    (b) (i) 35 hundredths   (ii) $0.35\,m^2$

    (c) (i) 64 hundredths   (ii) $0.64\,m^2$

A11 (a) $0.36\,m^2$ (b) $0.18\,m^2$ (c) $0.63\,m^2$

A12 (a) $0.9\,m$ (b) $0.6\,m$ (c) $0.8\,m$

A13 He is not right, $0.3 \times 0.2 = 0.06$.
The pupil's diagram should show a 10 by 10 grid with 6 small squares shaded.

A14 (a) $0.08\,m^2$ (b) $0.12\,m^2$ (c) $0.09\,m^2$

A15 (a) A = $1.00\,m^2$ B = $0.30\,m^2$
        C = $0.70\,m^2$ D = $0.21\,m^2$

    (b) $2.21\,m^2$

    (c) The calculator result should be the same, $2.21\,m^2$

A16 Whole square metres: $2\,m^2$
Pieces: two that are $0.8\,m^2$,
one that is $0.6\,m^2$
and one that is $0.48\,m^2$
making $4.68\,m^2$ in total.
Multiplying 2.6×1.8 on a calculator gives the same result.

A17 (a) $3.23\,m^2$ (b) $5.76\,m^2$ (c) $4.64\,m^2$

    (d) $5.4\,m^2$ (e) $7.35\,m^2$ (f) $9.8\,m^2$

*A18 (a) $2.5\,m$ (b) $0.75\,m$ (c) $0.5\,m$

A19 (a) 3.6 cm by 4.8 cm, $17.3\,cm^2$

    (b) 3.1 cm by 6.3 cm, $19.5\,cm^2$

    (c) 5.2 cm by 3.8 cm, $19.8\,cm^2$

Ⓑ Areas of other shapes (p 57)

B1 (a) The pupil's sketch

    (b) $65\,m^2$

B2 (a) $14\,cm^2$ (b) $\frac{1}{2}$ (c) $7\,cm^2$

B3 (a) $8.0\,cm^2$ (b) $5.6\,cm^2$ (c) $7.8\,cm^2$

B4 (a) $20\,m^2$ (b) $22\,m^2$ (c) $24\,m^2$

What progress have you made? (p 58)

1 $15.54\,m^2$

2 3.6 cm

3 $104\,m^2$

Practice booklet

Section A (p 29)

1 (a) $2.4\,m^2$ (b) $3.5\,m^2$ (c) $5.7\,m^2$

    (d) $12.71\,m^2$ (e) $25.2\,m^2$ (f) $4.1\,m^2$

2 The areas are 2.25 cm², 6.25 m², 12.25 cm², 20.25 cm².
In general, something-point-five squared equals the something squared plus one times the something with .25 added. This can be explained with a diagram like
this …

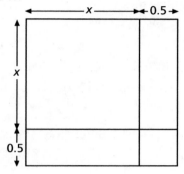

… which can be cut up and rearranged to make a diagram like this.

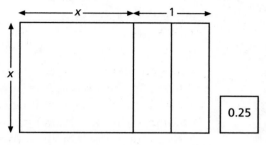

Section B (p 29)

1 Answers may differ from these slightly.

L 4.3 cm²	I 3.4 cm²	F 5.4 cm²
T 4.7 cm²	H 7.2 cm²	E 6.5 cm²

⑨ Finding rules

Pupils find expressions and formulas by analysing geometric patterns. They begin to see that, for a linear expression for the *n*th pattern, the coefficient of *n* is the difference between each successive pairs of terms.

Pupils are asked to find expressions *and* formulas throughout the unit. Where a formula is asked for, it is acceptable for pupils just to give an expression at this stage: writing their own formulas may be an unnecessary complication for some. Pupils should choose their own letter for the subject of any formula.

Practice booklet pages 30 to 33

Ⓐ **Maori patterns** (p 59)

◊ Ask the pupils (perhaps in pairs or groups) to count the crosses in each pattern shown on page 59. They could then continue the design by drawing further examples of this type.

Can pupils predict how many crosses will be in the 10th pattern? In the 100th? In the *n*th?

Get pupils to tell the rest of the class how they found the number of crosses in the 100th pattern.

◊ In discussion, try to bring out different ways of seeing how many crosses are in each pattern. For example:

• 'Pattern 10 will have 10 crosses to the right, 10 to the left and 10 at the bottom, plus 1 more in the middle. So that's $3 \times 10 + 1$ for the 10th pattern, or $3n + 1$ for the *n*th pattern.'

• 'Pattern 10 will have 10 crosses to the right, 10 to the left and 11 down the middle. So that's $10 + 10 + 11$ for the 10th pattern, or $n + n + n + 1 = 3n + 1$ for the *n*th pattern.'

◊ Discuss the difference between an **expression** for the number of crosses in the nth pattern $(3n + 1)$ and a **formula** $(c = 3n + 1)$.

◊ There is no need to bring out the link between 3 crosses being added each time and the 3 in front of the n at this stage. This will be developed in sections B and C.

A3, A4 The geometry of the design gives the opportunity to discuss equivalent expressions with pupils.

Examples for A3:

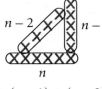   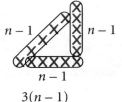

$$n + (n - 1) + (n - 2)$$ $$3(n - 1)$$

Examples for A4:

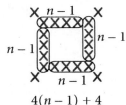   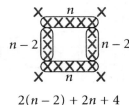   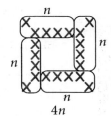

$$4(n - 1) + 4$$ $$2(n - 2) + 2n + 4$$ $$4n$$

B Matchstick patterns (p 61)

Pupils begin to look at differences in this section (the number of matches added to one pattern to make the next).

An additional activity is to give students a formula and ask them to design their own matchstick pattern to fit the formula.

C Finding formulas (p 62)

◊ Once pupils have tried to answer the questions on the page for each set of patterns, they can share their ideas in a whole-class discussion.

Demonstrating on an OHP may help pupils appreciate the relationship between the constant difference and the coefficient of n.

For examples, you could use matches to build up patterns in the matchstick design as below.

$$1 + \mathbf{3} \times 1$$ $$1 + \mathbf{3} \times 2$$ $$1 + \mathbf{3} \times 3$$

The number of matches added on each time is the number in bold and will be the coefficient of n in the general expression $(1 + \mathbf{3} \times n)$.

Pattern problems (p 64)

T

◊ Again pupils can consider the problem in pairs or groups and then share their ideas in a whole-class discussion.

Pupils have already met problems like this in the previous section and may have solved them using trial and improvement. By the end of the discussion pupils should see how to express a problem like this as a simple equation, $3n + 1 = 238$ in this case, and solve it to find the solution (pattern 79).

Ⓐ **Maori patterns** (p 59)

A1 (a) The pupil's sketches of patterns 4 and 5

(b) The pupil's explanation of what pattern 10 looks like

(c)

Pattern number	1	2	3	4	5
Number of crosses	5	9	13	17	21

(d) (i) 41   (ii) 201   (iii) 401

(e) $4n + 1$

A2 (a) The pupil's sketches of patterns 4 and 5

(b) The pupil's explanation of what pattern 10 looks like

(c)

Pattern number	1	2	3	4	5
Number of crosses	4	6	8	10	12

(d) 202

(e) $2n + 2$ or equivalent

(f) Pattern 27

A3 (a) The pupil's sketches of patterns 3 and 5

(b)

Pattern number	3	4	5	6
Number of crosses	6	9	12	15

(c) 27 with the pupil's explanation

(d) 297

(e) $3n - 3$ or equivalent

A4 (a) The pupil's sketches of patterns 4 and 6

(b) The pupil's explanation of what pattern 10 looks like

(c) Pattern 3:   12 crosses
Pattern 4:   16 crosses
Pattern 5:   20 crosses
Pattern 6:   24 crosses
Pattern 10:   40 crosses

(d) 400 crosses

(e) $c = 4n$ or equivalent

(f) Pattern 21

A5 The pupil's weaving pattern and formula

Ⓑ **Matchstick patterns** (p 61)

B1 (a) The pupil's sketch of pattern 4

(b)

Pattern number	1	2	3	4	5
Number of matches	2	4	6	8	10

(c) 2

(d) (i) 20   (ii) 200

(e) $m = 2n$

B2 (a) The pupil's sketch of pattern 4

(b) 2

(c) (i) 21   (ii) 201

(d) $m = 2n + 1$

B3 (a) The pupil's sketch of pattern 4

(b) 3

(c) 61

(d) $m = 3n + 1$

B4 (a) The pupil's sketch of pattern 4

(b) 5

(c) 502

(d) $m = 5n + 2$

ℂ Finding formulas (p 62)

C1 (a) 11 (b) 14 (c) 3
(d) $c = 3n + 2$ (e) 62

C2 (a) 4 (b) $c = 4n + 3$
(c) 43 (d) Pattern 20

C3 (a) $m = 5n + 1$ (b) $c = 251$
(c) Pattern 19

C4 (a) $m = 2n + 3$ (b) $m = 3n + 5$

C5 (a) $m = 6n - 2$ (b) 118
(c) Pattern 25

𝔻 Pattern problems (p 64)

D1 (a) 17 (b) 401 (c) $4n + 1$
(d) 181 (e) $4n + 1 = 349, \ n = 87$

D2 (a)

Pattern number	1	2	3	4	5
Number of beads	6	10	14	18	22

(b) $4n + 2$
(c) $4n + 2 = 470, \ n = 117$
(d) No, with the pupil's explanation

D3 (a) $5n + 3$
(b) $5n + 3 = 148, \ n = 29$
(c) Pattern 99, with the pupil's explanation

What progress have you made? (p 65)

1 (a) The pupil's sketches of patterns 3 and 5

(b)

Pattern number	2	3	4	5
Number of crosses	10	13	16	19

(c) 34 (d) $3n + 4$ (e) 304
(f) $3n + 4 = 163, \ n = 53$
(g) Yes, with the pupil's explanation

Practice booklet

Sections A and B (p 30)

1 (a) The pupil's sketch of patterns 4 and 5
(b) The pupil's explanation of pattern 10

(c)

Pattern number	1	2	3	4	5	6
Number of tiles	4	6	8	10	12	14

(d) 42 (e) $2n + 2$ (f) 202

2 (a) The pupil's sketch of patterns 3 and 5

(b)

Pattern number	2	3	4	5
Number of tiles	12	16	20	24

(c) The pupil's explanation of pattern 20; 84 tiles
(d) $4n + 4$ or equivalent
(e) Pattern 23

3 Design A
(a) The pupil's sketch of the next pattern
(b) 2 (c) $d = 2n + 1$

Design B
(a) The pupil's sketch of the next pattern
(b) 3 (c) $d = 3n + 1$

Design C
(a) The pupil's sketch of the next pattern
(b) 4 (c) $d = 4n + 4$

Sections C and D (p 32)

1 (a) Pattern 1: 8 beads
Pattern 2: 13 beads
Pattern 3: 18 beads
(b) 23 (c) 5
(d) $b = 5n + 3$ (e) 153

2 (a) 22 (b) 28
(c) 6 (d) $l = 6n + 4$
(e) 64

3 $d = 3n + 2$

4 (a) $l = 2n + 5$ (b) $l = 5n + 2$

5 (a) $2n + 1$ (b) 21
(c) (i) $2n + 1 = 241, \ n = 120$
(ii) 120

6 (a) $7n - 2$
(b) $7n - 2 = 40, \ n = 6$

10 Using a spreadsheet

 A number of activities relating to different topic areas are gathered together here. If your class has easy access to computers, then you could dip in whenever appropriate. If, however, you have access only at specific times, you could use the unit as a programme of work for your computer sessions. Alternatively, you could ask for the activities to be used in IT lessons which focus on the use of a spreadsheet.

> **Essential**
> Spreadsheet, or similar facility on a graphic calculator

Spot the formula (p 66)

◊ Some pupils may invent formulas that are impossibly difficult to spot.

◊ Opportunities to discuss equivalent formulas may arise, for example = (A1 + 3)*2 and = 2*A1 + 6.

◊ It is interesting to discuss strategies. For example, putting 100 or 1000 into a formula can often tell you a lot.

Making a sequence (p 67)

◊ You may need to introduce or revise the process of filling down a formula (or 'drag and drop').

◊ Pupils may find a formula for going from one term to the next. Although this is valid, get them to focus on finding a formula that works *across*, calculating the terms of the sequence from the numbers 1, 2, 3, …

Big, bigger, biggest (p 67)

The idea of using decimals may not occur at first.
Here are some solutions (others are possible in some cases):

1   (a) 2, 2, 3, 7   (b) 2, 3, 3, 6   (c) 3.5, 3.5, 3.5, 3.5 give 150.0625

2   (a) 3, 3, 4, 5   or   2.5, 4, 4, 4.5   (b) 2, 2.5, 5, 5.5

    (c) 3.75, 3.75, 3.75, 3.75 give 197.753 906 25

Ways and means (p 68)

Pupils will soon see that the sequence tends to a limit. The challenge is to find how the limit is related to the starting numbers. For two starting numbers a, b the limit is $\frac{1}{2}(a + 2b)$, for three starting numbers a, b, c it is $\frac{1}{6}(a + 2b + 3c)$, and for four, a, b, c, d it is $\frac{1}{10}(a + 2b + 3c + 4d)$.

Parcel volume (p 68)

The main focus here is on working systematically. As with many problems of this type it helps to restrict the number of variables. If the spreadsheet is set up in the most obvious way (as shown in the pupil's book), it may be difficult to keep track of the restrictions on the variables.

Ask pupils how they could build in the connection between length and girth (which must add up to 3 m for maximum volume). One way is to use a formula equivalent to '3 − girth' in the length column. As girth is already defined as 2(height + width), this reduces the variables to two: height and width.

The maximum volume of $0.25 \, \text{m}^3$ occurs when the width and height are both 0.5 m and the length is 1 m.

Furry festivals (p 69)

Pupils may themselves suggest reducing the number of variables by setting the number of pens equal to 60 minus the total number of T-shirts and badges.

The solution is:

    5 T-shirts, 35 badges and 20 pens

 or 6 T-shirts, 14 badges and 40 pens

Breakfast time (p 69)

The essential step is to set up a formula for working out a given percentage of a quantity. Pupils may know that, for example, 8 in the percentage column has to be converted to 0.08 but may not immediately realise that the operation needed to do this is ÷ 100.

Estimation

Pupils estimate a range of lengths.

The unit also gives practice in some easy conversion methods between common metric and imperial measurements.

p 70	**A** Good judgement	Estimating metric distances
p 71	**B** Converting lengths	Converting between metric and imperial lengths
p 73	**C** Liquid measures	Converting between metric and imperial capacity measures
p 74	**D** Weights	Converting between metric and imperial weights

Essential	**Optional**
Metre rules	Chalk
Tape measures	Plant markers
Trundle wheel	
Counters	
Local maps	
Practice booklet pages 34 and 35	

Ⓐ **Good judgement** (p 70)

Metre rules, tape measures, trundle wheel, counters, local maps
Optional: Plant markers, chalk

One objective of this section is to provide pupils with 'markers' on which they can base their estimates. Pupils may have some useful ones already from their own experiences; for example, swimmers will have a good idea what 25 m is, and most doors are about 2 m high. Pupils should be encouraged to use these and describe them to others.

Each activity asks pupils to confirm a distance by measurement and then use this to estimate other distances.

◊ **How long is one metre?** Pupils could start by suggesting a length in the classroom that is 1 m. Drapers often stretch out one arm and use the distance from their fingertips to the opposite shoulder to estimate a metre. Pupils could hold a piece of chalk or marker in each hand and try to mark out 1 m on the board, a large piece of paper or a wall.

◊ **How long is 100 m?** Pupils could mark their estimates with plant markers which are fairly inexpensive and reusable. After one estimate at 100 m, pupils could then estimate 100 m behind the line and see if estimates improve. Only a small number of trundle wheels is needed – it is useful if the class follow round whoever is measuring and watch their estimates being confirmed or otherwise. Tape measures could be used but are rather clumsy.

◊ **How far is one kilometre?** This activity will be enhanced by having a prepared list of local key places to which pupils can estimate the distance from the school. Checking against a local map is important.

Pupils could also consider the same problems, but estimate the distances in miles.

B **Converting lengths** (p 71)

◊ Some pupils may not know how tall they are in feet and inches but do in metres and centimetres, or vice versa. This is a good opportunity to convert from inches into centimetres.

Pupils could produce flow diagrams to convert from imperial to metric measurements and vice versa.

◊ Point out that, although these methods are reasonably accurate, they are not exact. For example 1 mile is equivalent to 1.609 km (to 4 s.f.) and not 1.6 km as the rule suggests. However, they will produce results that are good enough for most practical purposes.

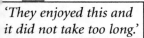

◊ Pupils could use a spreadsheet to produce ready reckoners for converting between metric and imperial measurements.

B12 A hand is officially 4 inches. When a horse's height is given as 16.2 hh this means 16 hands and 2 inches.

C **Liquid measures** (p 73)

◊ More able pupils can compare the two rules on the page. For example, the first rule gives 16 gallons as 72 litres. However, the second rule gives 72 litres as 16.2 gallons. Neither of the two rules give exact results and they are not the inverse of each other but they can still be used to give sensible estimates.

D **Weights** (p 74)

Ⓐ Good judgement (p 70)

A1 (a) metres   (b) centimetres

     (c) kilometres   (d) millimetres

     (e) metres   (f) millimetres

A2 The pupil's estimate

A3 The pupil's estimate

A4 The pupil's estimate (A typical car park space is about 2 m wide.)

A5 The pupil's estimate (The Skylon Tower is in fact 555 m tall.)

*A6 A VW Polo is about 3.5 m long. About 360 000 of these cars

Ⓑ Converting lengths (p 71)

B1 Calais   10   Boulogne 15

     Firenze   40   Siena   20

     Limerick   32.5/33   Ennis   12.5/13

B2 (a) 80 cm   (b) 100 cm   (c) 40 cm

B3 9 metres

B4 (a) False   (b) True   (c) True

B5 1.8 metres

B6 800 km

B7 (a) 48 km/h   (b) 80 km/h   (c) 112 km/h

B8 (a) 2 inches

     (b) (i) 6 inches   (ii) 28 inches

B9 5 feet

B10 3 inches

B11 (a) False   (b) False   (c) True

B12 (a) 150 cm   (b) 170 cm   (c) 160 cm

Ⓒ Liquid measures (p 73)

C1 (a) 36 litres   (b) 90 litres

     (c) 225 litres   (d) 900 litres

C2

Wayne	2 gallons	9 litres
Zak	6 gallons	27 litres
Irvine	5 gallons	22.5 litres
Stephen	30 gallons	135 litres

C3 (a)

Monday	7.2 gallons
Tuesday	5.4 gallons
Wednesday	6.3 gallons
Thursday	4.5 gallons
Friday	9.9 gallons
Saturday	8.1 gallons

     (b) 41.4 gallons

     (c) About 29 miles per gallon

C4 A pint is slightly more than a $\frac{1}{2}$ litre. The pupil's explanation such as: 'A gallon is 8 pints which is equivalent to 9 lots of a $\frac{1}{2}$ litre so a $\frac{1}{2}$ litre must be less than a pint.'

C5 A bit less than 2 pints

Ⓓ Weights (p 74)

D1 (a) 66 pounds   (b) 11 pounds

     (c) 77 pounds   (d) 17.6 pounds

D2 Tony and Jo

D3 Yes, 3 kg is equivalent to about 6.6 pounds which is slightly more than $6\frac{1}{2}$ pounds.

D4 500 grams is heavier. 500 grams is $\frac{1}{2}$ kg which is equivalent to about 1.1 pounds which is more than 1 pound.

What progress have you made? (p 74)

1 The pupil's estimate

2 60 miles is further.
80 km is equivalent to 50 miles.

3 70 cm

4 54 litres

5 7.7 kg

6 No, 1 litre is less than 2 pints.

Practice booklet

Sections A, B, C and D (p 34)

1 (a) The pupil's estimate of room length and width

(b) The area of the pupil's room

(c) Answer (b) × £25

2 (a) The pupil's estimate of a car's length in cubits (approximately 9 cubits)

(b) Answer (a) × $\frac{1}{2}$ in metres

3 (a) 5   (b) 5   (c) 3   (d) 1 inch

4
Bologna to Venice	100 miles
Florence to Genoa	150 miles
Rome to Naples	125 miles
Florence to Bologna	62.5 miles

5
Lincoln to Birmingham	144 km
York to Dundee	400 km
Leeds to Dover	416 km
Edinburgh to London	624 km

6
David	150 cm
Sarah	130 cm
Diana	122.5 cm
Tim	170 cm

7 Yes, 1.2 metres is about 4 feet.

8 (a) 2   (b) 4

9
Volvo	45 litres
Jeep	58.5 litres
Dart	216 litres (roughly 225 litres)

10 B

11 Yes, it contains over 2 pounds.

12 John

13 (a) grams   (b) millimetres

(c) litres   (d) metres

(e) kilograms   (f) centimetres

(g) kilometres   (h) millilitres

⑫ Quadrilaterals

The names and properties of the square, rectangle, parallelogram, rhombus, trapezium, 'kite' and 'arrowhead' are revised or introduced. Pupils see how special types of quadrilaterals can be made up from certain types of triangles. The sum of the interior angles of a quadrilateral is established and used to find missing angles. The fact that some types of quadrilateral are special cases of others is briefly explored.

Essential	**Optional**
Square dotty paper Sheet 164 (preferably on card) Scissors Angle measurers	OHP transparencies based on the diagrams at the top of page 82

Practice booklet pages 36 and 37 (needs square dotty paper and angle measurers)

𝔸 Special quadrilaterals (p 75)

◊   Here is an opportunity to find out through discussion how many of the quadrilaterals' names and properties pupils know and to fill in any gaps in their knowledge.

𝔹 Quadrilaterals from triangles (p 78)

> Sheet 164, scissors

Although this section takes time it provides consolidation: pupils have to recognise triangle types and the special quadrilaterals they produce. It also offers practice in exploring all possibilities and can help pupils visualise quadrilaterals as built up from triangles. The section is well suited to work in small groups.

◊ You may need to revise the different types of triangles before starting this section. Remember that some types of quadrilaterals are special cases of others (a point dealt with more fully later). So, for example, if a pupil creates a rhombus in B3 and labels it 'Parallelogram' that is not wrong, but you could ask 'What special kind of parallelogram?'

C Angles of a quadrilateral (p 79)

> Angle measurers
> Optional: OHP transparencies based on the diagrams at the top of page 82

◊ Here, although pupils are looking at a particular quadrilateral, the conclusion – that the sum of the interior angles of a quadrilateral is 360° – must be true for *every* quadrilateral. This is a deductive argument, an informal *proof*: if what we know about the angle sum of any triangle is true, our conclusion about quadrilaterals must be true. (Pupils discovered, but did not prove, the sum of the angles of a triangle in *Book 1*.) What pupils find from C1 has a different status: it is merely experimental confirmation (and rounding during measuring may give a total not quite equal to 360°).

◊ The green panel at the top of page 82 can be used as the basis for initial work on corresponding angles and alternate angles. The third bullet point suggests an informal proof that the sum of the angles of a triangle is 180°.

D Drawing and describing (p 82)

> Angle measurers

◊ In questions D4 to D7, some pupils may need reassurance that the sketches are not trying to indicate the *shape* of the quadrilaterals: they merely show how the vertices are lettered.

E Stand up if your drawing ... (p 83)

◊ A parallelogram can be defined as any quadrilateral that has two pairs of parallel sides. So a rectangle is a special kind of parallelogram.
Similarly,
a rhombus is a special kind of a parallelogram,
a square is both a special rhombus and a special rectangle,
a square is a special parallelogram,
and so on.

Introduce these ideas through discussion before going on to the following activity.

Six pupils sit on chairs facing the class each holding one of these drawings.

Say to the group 'Stand up if your drawing is a rectangle.' If the person with the square doesn't stand up, ask the class if there is anyone not standing up who should be and, by inviting explanations, check that the special case idea has been understood.

Repeat the process with 'Stand up if your drawing ...

     ... is a rhombus

     ... is a square'

and so on. You can also extend the idea to properties of their drawings: 'Stand up if your drawing ...

     ... has reflection symmetry

     ... has at least one right angle'

and so on.

After you have done these activities, pupils can decide what the teacher asked for in the photographs shown on the page.

Note that a trapezium is sometimes defined as having only one pair of parallel sides and sometimes as having at least one pair of parallel sides.

Ⓐ **Special quadrilaterals** (p 75)

A1 to A5 The pupil's drawings on dotty paper

A6 (a) P, Q, R and U are trapeziums.

    (b) P and U have one line of reflection symmetry.

Ⓑ **Quadrilaterals from triangles** (p 78)

B1 These quadrilaterals are possible.

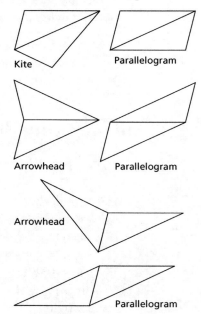

Kite

Parallelogram

Arrowhead

Parallelogram

Arrowhead

Parallelogram

B2 You could not make an arrowhead.

B3 These quadrilaterals are possible.

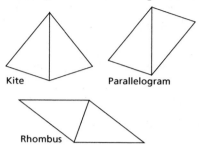

Kite   Parallelogram

Rhombus

B4 A kite would not be possible but an arrowhead would.

B5 Only this is possible.

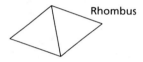

Rhombus

B6 (a) These quadrilaterals are possible.

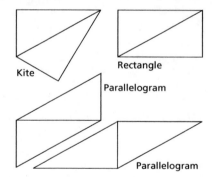

Kite   Rectangle

Parallelogram

Parallelogram

(b) These triangles are possible.

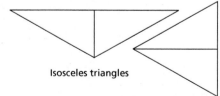

Isosceles triangles

B7 A square and only one other parallelogram would be possible. Only one isosceles triangle would be possible, a right-angled one.

B8 (a) $\frac{1}{2}$ (b) $\frac{1}{3}$ (c) $\frac{1}{4}$ (d) $\frac{1}{2}$

Ⓒ Angles of a quadrilateral (p 79)

C1 The pupil's drawing. The angles should add up to 360° (or a value very close).

C2 $a = 140°$, $b = 70°$, $c = 67°$

C3 $a = 40°$, $b = 97°$, $c = 116°$, $d = 258°$

C4 $a = 220°$, $b = 80°$, $c = 105°$, $d = 250°$, $e = 40°$, $f = 110°$, $g = 65°$, $h = 100°$, $i = 105°$, $j = 75°$, $k = 105°$, $l = 76°$, $m = 80°$, $n = 92°$, $o = 112°$

C5 $a = 100°$, $b = 112°$, $c = 115°$, $d = 65°$, $e = 225°$, $f = 45°$

C6 (a) A parallelogram

(b) A program of the following form, with $p = r = t = v$, $q = u$, $s = 180 - q$
FORWARD p
RIGHT q
FORWARD r
RIGHT s
FORWARD t
RIGHT u
FORWARD v

(c) A program of the above form, with $p = r = t = v$, $q = s = u = 90$

(d) A program of the above form, with $p = t$, $r = v$, $q = s = u = 90$

*C7 $a = 135°$, $b = 45°$, $c = 60°$, $d = 45°$, $e = 30°$

Ⓓ Drawing and describing (p 82)

D1 to D3 The pupil's drawings and checks

D4 (a) The pupil's drawing
(b) A rectangle

D5 (a) The pupil's drawing
(b) A rhombus

*D6 A kite

*D7 (a) A square   (b) A parallelogram
(c) A trapezium (a symmetrical or 'isosceles' one)

E **Stand up if your drawing ...** (p 83)

The teacher said:

1 'Stand up if your drawing is a parallelogram.'

2 'Stand up if your drawing has exactly two lines of reflection symmetry.'

3 'Stand up if your drawing has a pair of parallel sides, (or, possibly, '... if your shape is a quadrilateral').

What progress have you made? (p 84)

1 The pupil's drawings

2 A kite, an arrowhead (a trapezium might have)

3 A square

4 $a = 92°$, $b = 240°$, $c = 55°$

5 The pupil's drawings

6 Square, rhombus, rectangle

Practice booklet

Section A (p 36)

1 The pupil's drawings

Section B (p 36)

1 The pupil's drawings, for example,

(a) 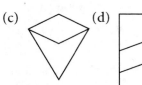 (b)

(c) (d)

(e)

Section C (p 36)

1 $a = 93°$, $b = 98°$, $c = 230°$

2 $a = 92°$, $b = 58°$, $c = 117°$, $d = 102°$, $e = 110°$, $f = 72°$, $g = 108°$, $h = 48°$, $i = 95°$, $j = 68°$, $k = 244°$

3 $a = 84°$, $b = 94°$, $c = 36°$

Section D (p 37)

1 The pupil's drawings

Negative numbers

Pupils revisit adding and subtracting and are introduced to multiplying and dividing negative numbers.

	Optional
	Sheet 192
	Counters, tiles or small cards of different colours
Practice booklet pages 38 to 40	

A Adding (p 85)

> Optional: Sheet 192 (blank walls)

◊ This is revision of work done in *Book S1* 'Negative numbers'. The examples could be done orally and discussed. You could give pupils some more similar simple additions to do until you feel they have properly recalled the earlier work.

B Subtracting (p 87)

> Optional: Counters, tiles or small cards of different colours

◊ This is also revision of work done in *Book S1* 'Negative numbers'; it can be used as suggested above.

The subtraction strips provide an opportunity for further consolidation.

Adding and subtracting by 'zapping'

◊ Manipulating negative numbers is notoriously difficult and some pupils may respond to an alternative 'justification'. The approach outlined below is essentially the same as that taken in *Book S1* 'Negative numbers' but set in a different context.

Pupils need a supply of counters, tiles or small cards of two different colours. Each represents $^+1$ or $^-1$ (depending on colour) and can be labelled as such. In the examples below, a black counter represents $^+1$ and a white counter represents $^-1$.

Some pupils may like to think of the counters as units of 'matter' and 'anti-matter'. The key point is that a $^+1$ 'cancels' or 'zaps' a $^-1$ to give 0. Hence numbers can be represented in different ways.

For example, each of these sets of counters represent $^+2$.

An example of adding is:

4 + ⁻1 = 3   ● ● ● ● + ○ = ● ● ● ● / ○ = ● ● ●

For subtraction, it is sometimes necessary to add some zeros (pairs of black and white counters) in order to complete the subtraction.

An example is:

⁻2 – ⁻5 = 3   ○ ○   is equivalent to   ○ ○ ○ ○ ○ / ● ● ●   and now five white

counters can be subtracted (to represent subtracting $^-5$) to leave three black counters that represent $^+3$ or just 3.

One advantage of this approach is that subtractions such as $^-5 - ^-2$ become 'easy'. A disadvantage is that subtractions such as $2 - 5$ that are easy to visualise on the number line become harder.

ℂ **Multiplying** (p 88)

> Optional: Sheet 192 (blank walls)

◊ The first box shows how the product of a positive and a negative can be interpreted as repeated addition.

◊ In the second box the first multiplication table extends in an obvious way. The second table, going backwards, leads to the multiplication of two negative numbers. To preserve the pattern of adding 3s, the results have to be positive. It may be helpful to give the pupils some more similar multiplication patterns to continue.

◊ Finally the multiplication square can be copied and filled in.

D Dividing (p 90)

D5 You may wish to warn that some chains are quite long, for example 14 and 17.

E Mixed problems (p 91)

A Adding (p 85)

A1 (a) (i) $^-2 + 3 + 1 = 2$

(ii) $1 + {}^-2 + 3 + {}^-7 = {}^-5$

(b) The complete list is

$^-9 = {}^-2 + {}^-7$

$^-8 = {}^-2 + {}^-7 + 1$

$^-6 = 1 + {}^-7 = {}^-2 + {}^-7 + 3$

$^-5 = {}^-2 + {}^-7 + 3 + 1$

$^-4 = 3 + {}^-7$

$^-3 = 1 + 3 + {}^-7$

$^-1 = 1 + {}^-2$

$1 = {}^-2 + 3$

$2 = 3 + 1 + {}^-2$

$3 = {}^-2 + {}^-7 + 12$

$4 = 3 + 1 = 1 + {}^-2 + {}^-7 + 12$

$5 = {}^-7 + 12$

$6 = 1 + {}^-7 + 12 = {}^-2 + 3 + {}^-7 + 12$

$7 = {}^-2 + 3 + {}^-7 + 12 + 1$

$8 = 3 + 12 + {}^-7$

$9 = 1 + 3 + {}^-7 + 12$

$10 = {}^-2 + 12$

$11 = {}^-2 + 1 + 12$

$13 = 1 + 12 = {}^-2 + 3 + 12$

$14 = 1 + {}^-2 + 3 + 12$

$15 = 3 + 12$

$16 = 1 + 3 + 12$

A2 (a)

+	⁻3	5
2	**⁻1**	**7**
⁻1	**⁻4**	**4**

(b)

+	6	⁻1
⁻4	**2**	**⁻5**
6	**12**	**5**

(c)

+	12	⁻8
⁻7	**5**	**⁻15**
10	**22**	**2**

(d)

+	35	⁻19
13	**48**	**⁻6**
⁻3	32	⁻22

A3

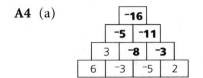

A4 (a)

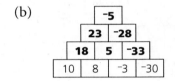

(b)

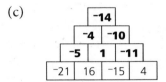

(c)

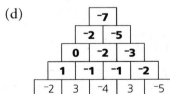

(d)

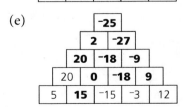

(e)

(f)

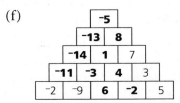

A5 (a)
$$^-2$$
$$1 \quad ^-3$$
$$3 - ^-2 - ^-1$$

(b) $4 - 2.8 - ^-1.2$
$4.3 \quad ^-0.9$
0.3

(c)
$$4$$
$$2 \quad 1$$
$$^-2 - ^-5 - ^-3$$

(d)
$$1$$
$$^-2 \quad 3$$
$$^-3 - ^-1 - 2$$

(e) $^-7 - ^-6 - 1$
$$^-4 \quad 4$$
$$3$$

(f)
$$4$$
$$^-3 \quad 11$$
$$^-7 - 0 - 7$$

B Subtracting (p 87)

B1 (a) | 16 | 14 | **2** | **12** | **-10** |

(b) | 13 | 2 | **11** | **-9** | **20** |

(c) | 8 | 9 | **-1** | **10** | **-11** |

(d) | 10 | -5 | **15** | **-20** | **35** |

(e) | -6 | 3 | **-9** | **12** | **-21** |

(f) | -2 | -5 | **3** | **-8** | **11** |

B2 (a) | 11 | 9 | 2 | 7 | **-5** |

(b) | **6** | **4** | **2** | **2** | **0** |

(c) | **3** | 5 | -2 | 7 | **-9** |

(d) | **10** | **7** | **3** | **4** | **-1** |

(e) | **5** | 2 | 3 | -1 | 4 |

(f) | **-5** | -2 | **-3** | 1 | **-4** |

B3 (a) (i) $^-1 - ^-3 - 2 = 0$ or
$^-1 - 2 - ^-3 = 0$

(ii) $^-3 - 2 - ^-1 = ^-4$

(b) The largest answer is 6 ($2 - ^-3 - ^-1$).

C Multiplying (p 88)

C1 (a) $^-25$ (b) $^-30$ (c) 30 (d) $^-20$
(e) $^-18$ (f) $^-18$ (g) $^-18$ (h) 18

C2 (a) 4 (b) 16 (c) 30 (d) $^-12$
(e) $^-36$ (f) $^-36$ (g) $^-180$ (h) $^-8$

C3 (a)
×	2	4
3	**6**	**12**
-2	**-4**	**-8**

(b)
×	5	-3
2	**10**	**-6**
-1	**-5**	**3**

(c)
×	-1	-8
-3	**3**	**24**
-4	**4**	**32**

(d)
×	-6	-7
3	**-18**	**-21**
-7	**42**	**49**

C4 (a) (i) $^-4 \times 1 = ^-4$ or $^-2 \times 2 = ^-4$

(ii) $^-4 \times ^-3 = 12$

(iii) $3 \times ^-2 \times ^-4 = 24$ or
$^-3 \times ^-4 \times 2 = 24$

(iv) $^-1 \times ^-4 \times ^-2 = ^-8$ or
$1 \times 2 \times ^-4 = ^-8$

(b) The smallest answer is
$^-24$ ($^-3 \times ^-2 \times ^-4$).

C5

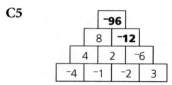

C6 (a)

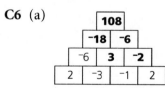

(b)

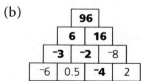

(c)

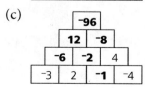

C7 (a) | 1 | -2 | -2 | 4 | -8 |

(b) | 2 | -2 | -4 | 8 | -32 |

(c) | -3 | 1 | -3 | -3 | 9 |

(d) | -1 | 3 | -3 | -9 | 27 |

(e) | -2 | -2 | 4 | -8 | -32 |

(f) | -5 | 2 | -10 | -20 | 200 |

*C8 (a) | 2 | -1 | -2 | 2 | -4 |

(b) | -1 | 2 | -2 | -4 | 8 |

(c) | 3 | -2 | -6 | 12 | -72 |

D Dividing (p 90)

D1
(a) $-15 \div 3 = -5$ $-15 \div -5 = 3$

(b) $21 \div -7 = -3$ $21 \div -3 = -7$

(c) $-12 \div 4 = -3$ $-12 \div -3 = 4$

(d) $-16 \div -2 = 8$ $-16 \div 8 = -2$

(e) $20 \div -4 = -5$ $20 \div -5 = -4$

(f) $-14 \div -2 = 7$ $-14 \div 7 = -2$

(g) $-14 \div 2 = -7$ $-14 \div -7 = 2$

(h) $14 \div -2 = -7$ $14 \div -7 = -2$

D2 pos $\div$ pos = positive
neg $\div$ pos = **negative**
pos $\div$ neg = **negative**
neg $\div$ neg = **positive**

D3 (a) 4 (b) 10 (c) 4 (d) -5

(e) 8 (f) 4 (g) -7 (h) -5

(i) -7 (j) -10 (k) 30 (l) -18

D4 (a) -5 (b) -6 (c) 7 (d) -7

(e) -4 (f) 4 (g) -9 (h) -12

(i) -3 (j) -4 (k) -1.5 (l) 0.5

D5 $6 \to {}^-3 \to 10 \to {}^-5 \to 16 \to {}^-8 \to$
$4 \to {}^-2 \to 1 \to {}^-2$
The pupil's own chains, always ending in the same 'cycle' (so long as the starting number is not 0)

E Mixed problems (p 91)

E1 (a) -5 (b) -11 (c) 7 (d) -10

(e) 12 (f) 4 (g) -6 (h) 2

E2 (a) 4 (b) -5 (c) 8 (d) -23

(e) -27 (f) -8 (g) 13 (h) -2

E3 (a) -5 (b) -5 (c) 12 (d) 14

E4 (a) -8 (b) 12 (c) 4 (d) -4

(e) 3 (f) -6 (g) -2 (h) 40

E5 (a)

+	-2	3
8	6	11
5	3	8

(b)

+	1	6
-4	-3	2
2	3	8

(c)

+	-10	4
-5	-15	-1
-3	-13	1

(d)

+	-12	-6
9	-3	3
3	-9	-3

E6 (a) | 21 | -9 | 12 | 3 | 15 |

(b) | 41 | -24 | 17 | -7 | 10 |

(c) | 12 | -5 | 7 | 2 | 9 |

E7 (a)

×	-2	-5
3	-6	-15
-4	8	20

(b)

×	2	-3
3	6	-9
-6	-12	18

(c)

×	10	-5
-2	-20	10
-3	-30	15

(d)

×	1	-0.2
-1	-1	0.2
2.5	2.5	-0.5

E8 The pupil's own calculations

Get to 1 or ⁻1

Each number can lead to 1 or ⁻1 in two stages. Examples are

$^-2$ —[× 3]—[÷ ⁻6]→ 1
3 —[× ⁻2]—[÷ ⁻6]→ 1
$^-4$ —[− ⁻6]—[÷ ⁻2]→ ⁻1
5 —[− 3]—[÷ ⁻2]→ ⁻1
$^-6$ —[− ⁻4]—[÷ ⁻2]→ 1

What progress have you made? (p 92)

1 (a) 3   (b) 12   (c) ⁻3

2 (a) ⁻24 (b) 27   (c) ⁻50 (d) ⁻5
  (e) 4   (f) ⁻6   (g) ⁻3   (h) ⁻4
  (i) 5

3 (a) 12   (b) ⁻15   (c) 10   (d) 5
  (e) 3   (f) ⁻6

4 (a) 2   (b) ⁻12   (c) 12   (d) 4

Practice booklet

Sections A and B (p 38)

1 (a) 1   (b) ⁻3   (c) ⁻6   (d) ⁻2
  (e) ⁻4   (f) ⁻7   (g) ⁻11 (h) ⁻12

2 (a)
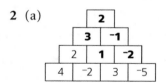

(b)

(c)
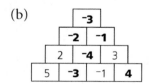

3 (a) 2   (b) ⁻6   (c) ⁻6   (d) 7
  (e) 11   (f) 1   (g) ⁻1   (h) 2

4 (a)

| 10 | 8 | 2 | 6 | ⁻4 |

(b)

| 5 | 7 | ⁻2 | 9 | ⁻11 |

(c)

| 8 | ⁻3 | 11 | ⁻14 | 25 |

(d)

| ⁻4 | ⁻2 | ⁻2 | 0 | ⁻2 |

(e)

| 12 | 17 | ⁻5 | 22 | ⁻27 |

(f)

| ⁻5 | ⁻3 | ⁻2 | ⁻1 | ⁻1 |

5 (a) $^-4 + {}^-2 - 1 = {}^-7$
  (b) $^-2 - {}^-4 - 1 = 1$
  (c) $1 + {}^-4 - {}^-2 = {}^-1$
  (d) $1 - {}^-4 - {}^-2 = 7$

Sections C and D (p 39)

1 (a)

×	3	5
2	6	10
⁻1	⁻3	⁻5

(b)

×	4	⁻3
⁻2	⁻8	6
⁻5	⁻20	15

(c)

×	⁻2	⁻1
6	⁻12	⁻6
⁻4	8	4

(d)

×	⁻7	⁻5
5	⁻35	⁻25
⁻4	28	20

2 (a) 24   (b) ⁻27   (c) ⁻30   (d) 20

3 (a)

		48		
	⁻4	⁻12		
	⁻2	2	⁻6	
2	⁻1	⁻2	3	

(b)

		⁻12		
	⁻4	3		
	⁻4	1	3	
4	⁻1	⁻1	⁻3	

(c)

		16		
	2	8		
	⁻1	⁻2	⁻4	
⁻1	1	⁻2	2	

4 (a) $12 \div {}^-3 = {}^-4$, $^-12 \div 3 = {}^-4$
  (b) $12 \div 4 = 3$, $\quad^-12 \div {}^-4 = 3$
  (c) $12 \div 3 = 4$, $\quad^-12 \div {}^-3 = 4$
  (d) $12 \div {}^-4 = {}^-3$, $^-12 \div 4 = {}^-3$

5 (a) ⁻8   (b) 5   (c) ⁻6   (d) 7

6 (a) ⁻12   (b) ⁻4   (c) 42   (d) ⁻5
  (e) ⁻36   (f) 4   (g) 24   (h) ⁻4
  (i) ⁻35   (j) 3   (k) ⁻54   (l) ⁻4

Section E (p 40)

1 (a) ⁻5   (b) ⁻14   (c) 7   (d) 5
  (e) 15   (f) ⁻12   (g) 8   (h) ⁻8

2 (a) 1   (b) 4   (c) ⁻4   (d) 60
  (e) ⁻1   (f) ⁻7   (g) 19   (h) 28

3 (a) ⁻2   (b) ⁻11   (c) 9   (d) 20

4 (a) ⁻10   (b) 2   (c) ⁻2   (d) 1.5
  (e) 16   (f) 3   (g) 28   (h) ⁻28

5 (a)

⁻9	4	⁻5	⁻1	⁻6

(b)

18	⁻14	4	⁻10	⁻6

(c)

3	⁻4	⁻1	⁻5	⁻6

6 (a) $\dfrac{3 - ^-3}{^-2} = ^-3$, $\dfrac{^-3 - 3}{2} = ^-3$, $\dfrac{2 - ^-4}{^-2} = ^-3$

  (b) $^-2(2 + ^-4) = 4$, $2(^-2 + 4) = 4$

  (c) $\dfrac{^-4}{3 - 2} = ^-4$, $\dfrac{4}{^-3 - ^-2} = ^-4$, $\dfrac{^-4}{4 - 3} = ^-4$,
      $\dfrac{4}{^-4 - ^-3} = ^-4$

Percentage change

> **Essential**
> Sheet 212
> **Practice booklet** pages 41 to 43

Ⓐ Increasing and decreasing (p 93)

The objectives here are to revise the equivalence between simple fractions and percentages and to use this to carry out percentage increases and decreases. The idea of using multipliers is introduced in subsequent sections.

> Sheet 212

◊ Start by doing some oral work on simple percentages, for example, 'What is 25% of 80 grams?' Then move on to the idea of increasing and decreasing by 25% by adding on or subtracting 25%.

Ⓑ Percentage increases and their multipliers (p 95)

◊ Pupils could decide on the correct multiplier in pairs and try to justify their decisions.

Ⓒ Percentage decreases and their multipliers (p 96)

◊ Again, pupils could decide on the correct multiplier in pairs and try to justify their decisions.

Ⓓ Mixed questions (p 97)

Ⓐ Increasing and decreasing (p 93)

A1 (a) 10 kg   (b) £45
     (c) 70 cm   (d) 20 ml

A2 (a) 25%   (b) 20%

A3 (a) 50%   (b) 10%

A4 (a) 66 kg   (b) £9

A5 (a) 44   (b) 90

A6 £240

A7 Puzzle 1

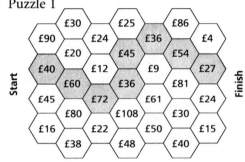

Puzzle 2

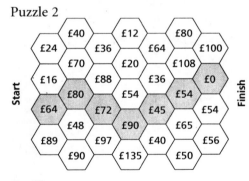

Puzzle 3

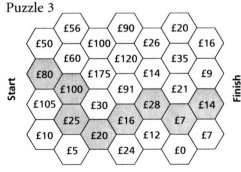

*A8 (a) £32   (b) £50

Ⓑ Percentage increases and their multipliers (p 95)

B1 (a) £35   (b) £6   (c) £71.25
     (d) £26.25   (e) £8.25

B2 (a) × 1.12   (b) £33.60

B3 (a) × 1.23   (b) £9.84

B4 (a) £18.40   (b) £28.75

B5 (a) £89.70   (b) 73 kg
     (c) 28.5 cm   (d) £20.76

B6 (a) × 1.05   (b) £1.68

Ⓒ Percentage decreases and their multipliers (p 96)

C1 (a) £12.40   (b) £23.56   (c) £44.64
     (d) £42.47   (e) £30.69

C2 A × 0.31   B × 0.85   C × 0.15
     D × 0.77   E × 0.6   F × 0.96

C3 (a) £25.50   (b) £64.60   (c) £4.59
     (d) £83.81   (e) £23.12   (f) £199.75

C4 (a) 0.72   (b) £17.64

C5 (a) £2.20   (b) £5.59
     (c) 259 kg   (d) 53.25 kg

C6 (a) 0.92   (b) 82.8 kg

Ⓓ Mixed questions (p 97)

D1 40.7 kg

D2 £30

D3 A × 0.81   B × 1.11   C × 0.66
     D × 1.8   E × 1.43   F × 0.19

D4 £4536

D5 22 200

D6 (a) £113.60   (b) 16.25 kg
     (c) 3948   (d) 53.1 kg

What progress have you made? (p 97)

1 (a) £15   (b) £40

2 £17.10

3 £239.20

Practice booklet

Section A (p 41)

1 (a) 12 kg   (b) £15
   (c) 33 cm   (d) 33 ml

2 (a) 25%   (b) 100%
   (c) 20%   (d) 10%

3 (a) 110 kg   (b) £24
   (c) 11   (d) 540

*4 (a) £50   (b) £80

Section B (p 42)

1 (a) × 1.37   (b) £95.90

2 (a) £11.90   (b) 32 kg
   (c) 35.76 m   (d) £118.30

3 88.5 kg

4 (a) £5.60   (b) £11.76

5 (a) £27.20   (b) 113.1 kg
   (c) 6.44 m   (d) £325.12
   (e) 69.6 kg   (f) 26.5 m

Section C (p 42)

1 (a) × 0.63   (b) £44.10

2 (a) £26.60   (b) 27.3 kg
   (c) 38.8 m   (d) £16.53

3 (a) £16.75   (b) £55.61
   (c) £12.06   (d) £4.49
   (e) £26.73   (f) £56.68

Section D (p 43)

1 A   × 0.94   B   × 1.02
   C   × 0.58   D   × 1.2
   E   × 1.58   F   × 0.4

2 (a) £102.20   (b) £129
   (c) 55.2 kg   (d) 54 kg
   (e) 884   (f) 9810

3 £8075

Review 2 (p 98)

1 (a) The pupil's sketches of patterns 4 and 5

  (b) 4   (c) 44

  (d) $c = 4n + 4$ or $c = 4(n + 1)$, where c is the number of counters

  (e) Pattern 15   (f) 36

2 8.64 m^2

3 (a) Rectangle or rhombus

  (b) Rhombus or rectangle

4 (a) $^-10$ (b) $^-1$ (c) 10 (d) 24

  (e) $^-5$ (f) $^-9$ (g) $^-3$ (h) $^-7$

5 (a) £48   (b) £12

6 0.4 m^2

7 (a) Probably 1.5 metres

  (b) About 2000

8 (a) The pupil's sketch of rhombus

  (b) Isosceles

9 4.8 cm^2

10 (a) $c = 5n + 3$ (b) 503

11 $a = 100°$, $b = 65°$, $c = 110°$, $d = 55°$

12 (a) 81 (b) 20 (c) 2 (d) $^-2$

13 £17

14 (a) The pupil's accurate drawings

  (b) (i) 130°   (ii) 117°

  (c) (i) Trapezium

     (ii) Parallelogram

15 (a) 5 kg is heavier as it is $5 × 2.2 = 11$ lb

  (b) 100 miles   (c) 20 cm

16 (a) $^-21$ (b) 2 (c) 34 (d) 20

17 108 members

Mixed questions 2 (Practice booklet p 44)

1 30 mph $≈$ 48 km/h hence 30 mph is slower.

2 2646

3 (a) Approximately 20 cm   (b) 10

4 (a) 31   (b) $3n + 1$

  (c) Pattern 33

5 (a) 11.97 m^2 (b) 7.35 m^2

  (c) $6.66 + 3.08 + 0.72 = 10.46$ m^2

6 (a) 22.5 litres   (b) 65 cm

7 DIAGONAL ($^-6$, $^-4$, $^-3$, $^-1$, 3, 4, 10, 24)

8 (a) $c = 6n - 4$ (b) 116

  (c) 15th pattern

9 (a) The pupil's accurate drawing of PQRS

  (b) The pupil's accurate drawing of TUVW

  (c) (i) QR = 8 cm

     (ii) VW = 6.1 cm

10 $a = 80°$ $b = 50°$ $c = 132°$

  $d = 72°$ $e = 108°$

11 (a) 2.8 m   (b) 8 cm

12 (a) $5 - {}^-4 + {}^-2 = 7$ or $^-2 - {}^-4 + 5 = 7$

  (b) $^-4 - {}^-2 + 5 = 3$ or $5 - {}^-2 + {}^-4 = 3$

  (c) $^-4 - 5 + {}^-2 = {}^-11$ or $^-2 - 5 + {}^-4 = {}^-11$

13 £79.50

14 (a) Parallelogram, kite or arrowhead

  (b) Kite

15 £1170

Probability from experiments

Essential	**Optional**
Plastic spoons (one per pair of pupils)	OHP and transparency of sheet 200
Sheet 200	Multilink cubes (10 per pair, see below)
Dice	Drawing pins (one per pair, see below)
Practice booklet pages 47 and 48	

A Experiments (p 100)

> Plastic spoons (one per pair of pupils), sheet 200
> Optional: multilink cubes (10 per pair), drawing pins (one per pair)

◊ You could start by holding up a plastic spoon and asking what will happen if it is dropped. Make it clear that although we might have a guess at the probability of it landing the right way up, the spoon does not have the right symmetry for us to use 'equally likely outcomes'.

The 0 to 1 scale is a way of thinking about probability. You could mark some of the class's guesses on a large scale on the board.

In one school the teacher had half the class doing the spoon experiment and the other half the drawing pin experiment.

◊ When pupils (in pairs) do their own trials, keep the recording informal for the first set of trials and collect together the class's results. (The complexity of the resource sheet might be counter-productive at first.)

When using the sheet, explain the first 'block' (recording the outcomes) first. When this has been completed, go on to explain the next block of columns, then the relative frequency, and finally the graph.

The graph should show how the relative frequency settles down to a value which can be used as an estimate of the probability.

◊ If everybody has used identical spoons, the class's results can be pooled to get a better estimate.

Dropping a drawing pin

Some pupils could do a comparative experiment, for example comparing two different surfaces. Sheet 200 could be used.

Dropping a multilink cube

The design of a suitable data collecting sheet can be left to the pupils.

Ⓑ **Relative frequency** (p 102)

◊ You may need to revise changing fractions to decimals using a calculator.

You can use these questions to draw attention to the fact that if the probability of an event occurring is p, the probability of it not occurring is $(1 - p)$. B6 requires pupils to know this.

Ⓒ **How often?** (p 104)

◊ This is informal work on the concept of 'expectation' in probability theory.

Ⓑ **Relative frequency** (p 102)

B1 $\frac{16}{50}$ or $\frac{8}{25}$ or 0.32

B2 (a) $\frac{12}{40}$ or $\frac{3}{10}$ or 0.3
(b) $\frac{18}{40}$ or $\frac{9}{20}$ or 0.45
(c) $\frac{10}{40}$ or $\frac{1}{4}$ or 0.25
The three probabilities add to 1.

B3 (a) $\frac{32}{50}$ or $\frac{16}{25}$ or 0.64
(b) $\frac{18}{50}$ or $\frac{9}{25}$ or 0.36
The coin does seem to be unfair.

B4 (a) Fred counted 124 papers altogether.
Mirror 0.18   *Sun* 0.15
Express 0.10   *Mail* 0.14
Star 0.09   *Telegraph* 0.12
Times 0.09   *Guardian* 0.08
Independent 0.06
They add up to 1.01 because of rounding.
(b) 0.15

B5 (a) $\frac{62}{250}$ or 0.248   (b) $\frac{44}{250}$ or 0.176

B6 (a) 0.048   (b) 0.952

Ⓒ **How often?** (p 104)

C1 (a) $\frac{1}{2}$ (b) 130

C2 (a) $\frac{1}{6}$ (b) 50
(c) (i) 50   (ii) 150   (iii) 250

C3 (a) 60   (b) 30   (c) 60

C4 (a) $\frac{40}{100}$ or $\frac{2}{5}$ or 0.4
(b) about 160 times

C5 (a) $\frac{44}{100}$ or $\frac{11}{25}$ or 0.44
(b) About 110 times

C6 About 350 times

C7 (a) 0.38   (b) About 91 times

What progress have you made? (p 105)

1 $\frac{12}{50} = \frac{6}{25} = 0.24$

2 0.34

3 About 80 times

4 About 1550

Practice booklet

Section B (p 47)

1 0.85 or $\frac{17}{20}$

2 (a) $\frac{10}{30}$ or $\frac{1}{3}$ (b) $\frac{11}{30}$ (c) $\frac{9}{30}$ or $\frac{3}{10}$

3 (a) $\frac{21}{40}$ (b) $\frac{19}{40}$

    Naomi seems better, but it is very close.

4 Jasmine counted 91 sales altogether.
   Ready salted 0.13
   Salt and vinegar 0.20
   Cheese and onion 0.23
   Prawn cocktail 0.25
   Beef 0.08
   Pickled onion 0.11

Section C (p 48)

1 (a) About 40 times
   (b) About 20 times
   (c) About 40 times

2 (a) About 83 times
   (b) About 250 times
   (c) About 167 times
   (d) About 167 times
   (e) About 417 times

3 (a) $\frac{23}{70}$ or approximately 0.33
   (b) About 66 times

4 (a) $\frac{83}{235}$ or 0.35 to 2 d.p.
   (b) Ready salted 353
      Salt and vinegar 221
      Cheese and onion 174
      Prawn cocktail 115
      Beef 81
      Pickled onion 55

Squares, cubes and roots

> **Practice booklet** pages 49 and 50

Ⓐ **Squares and cubes** (p 106)

T

◊ Review the ideas and vocabulary of square and cube numbers. Include the use of index notation for squares and cubes in your discussion.

A11/A14 These can be solved by trial and improvement. There is no need to introduce the square or cube root functions on a calculator at this stage.

Ⓑ **Roots** (p 107)

B9 to B11 These are intended to be solved by trial and improvement. There is no need to introduce the cube root function on a calculator at this stage.

Ⓒ **Using graphs** (p 108)

The graph helps to establish the inverse link between squaring and finding the square root and allows pupils to estimate the values of non-exact square roots.

It can also stimulate work on trial and improvement to find square (and cube) roots. Trial and improvement is covered in some depth in *Book S3* and this work is intended only as an introduction.

◊ Pupils can estimate some squares and square roots to start with.

Then you could pose a problem such as 'What is the length of the edge of a square that has an area of $10 \, \text{cm}^2$?' First establish that the side must be between 3 cm and 4 cm. Then ask for a suggestion to try out. Working with one decimal place, you find that the side must be between 3.1 cm and 3.2 cm. Now work with two decimal places, and beyond if the pupils wish to.

They should find that the side is between 3.16 cm and 3.17 cm so the exact value begins 3.16… which gives 3.2, correct to 1 d.p.

Working with three decimal places they should find that the side is between 3.162 cm and 3.163 cm so the exact value begins 3.162… which gives 3.16, correct to 2 d.p.

◊ Pupils could draw a graph of cube numbers and use it in the same way.

C4 It is intended that pupils solve part (d) by trial and improvement. However, you may wish to show some pupils how to use their calculators to find the cube root directly (if they have this facility).

𝔸 **Squares and cubes** (p 106)

A1 11

A2 4 + 16 = 20

A3 (a) 25 (b) 36 (c) 9 (d) 49

A4 64

A5 (a) 8 (b) 64 (c) 125 (d) 1

A6 1 + 8 = 9

A7 25 + 27 = 52

A8 (a) 9 − 4 = 5
(b) 9 − 1 = 8
(c) 36 − 25 = 11
(d) 25 − 1 = 24 or 49 − 25 = 24
(e) 49 − 36 = 13

A9 100, 121, 144

A10 512

A11 (a) 13 (b) 6 (c) 22 (d) 11

A12 1331

A13 (a) 1.44 (b) 42.875
(c) 0.81 (d) 0.125

A14 (a) 1.6 (b) 1.3 (c) 3.8 (d) 0.4

A15 (a) 4 (b) ⁻8 (c) 9 (d) ⁻27

A16 ⁻4

𝔹 **Roots** (p 107)

B1 (a) 16 (b) 16 (c) 25 (d) 25

B2 (a) 441
(b) (i) 21 (ii) ⁻21

B3 10

B4 ⁻5

B5 ⁻8, 8

B6 ⁻25

B7 (a) ⁻7, 7 (b) ⁻2, 2 (c) ⁻1, 1
(d) ⁻9, 9

B8 (a) 2 (b) 3 (c) 1
(d) 4 (e) ⁻3

B9 0.5

B10 (a) 9 cm (b) 9

B11 (a) 8 (b) 13 (c) 20
(d) 1.5 (e) ⁻12

ℂ **Using graphs** (p 108)

C1 (a) 12 (b) 2 (c) 6.8

C2 (a) 2.8 (b) ⁻3.7

C3 (a) 19.36
(b) 20.25
(c) 4.4^2 is less than 20 and 4.5^2 is greater than 20 so the square root of 20 must be between 4.4 and 4.5.
(d) 4.47
(e) ⁻4.47

C4 (a) 97.336

(b) 103.823

(c) 4.6^3 is less than 100 and 4.7^3 is greater than 100 so the cube root of 100 must be between 4.6 and 4.7.

(d) 4.64

What progress have you made? (p 109)

1 (a) 64   (b) 8

2 16

3 $1 + 27 = 28$

4 (a) 25   (b) $^-1$

5 (a) 7   (b) $^-9$ (c) 3   (d) $^-3$

6 (a) $^-3.9, 3.9$

(b) The pupil's explanation

Practice booklet

Sections A and B (p 49)

1 (a) 49   (b) 16   (c) 27   (d) 1

2 $64 + 8 = 72$

3 6 by 6 and 4 by 4

4 1 and 64 (and 729 and so on)

5 (a) 16   (b) $^-64$ (c) 25   (d) $^-125$

6 (a) 2025

(b) (i) 45   (ii) $^-45$

7 9

8 $^-12$

9 (a) 8, $^-8$ (b) 6, $^-6$

(c) 10, $^-10$ (d) 7, $^-7$

10 4 cubed is $4^3 = 4 \times 4 \times 4 = \mathbf{64}$
so the cube root of **64** is 4.
$(^-4)$ cubed is $(^-4)^3 = {}^-\mathbf{4} \times {}^-\mathbf{4} \times {}^-\mathbf{4} = {}^-\mathbf{64}$
so the cube root of $^-\mathbf{64}$ is $^-4$.

11 (a) $^-1$ (b) 5   (c) $^-2$ (d) 10

12 729 and 784

13 (a) 7.29   (b) 68.921

(c) 0.36   (d) 5.832

14 (a) 2.8   (b) 1.7   (c) 4.3   (d) 0.8

15 (a) 9   (b) 14   (c) 0.6   (d) $^-7$

Section C (p 50)

1 (a) 29.16

(b) 30.25

(c) 5.4^2 is less than 30 and 5.5^2 is greater than 30, so the square root of 30 must be between 5.4 and 5.5.

(d) 5.48

(e) $^-5.48$

2 (a) 195.112

(b) 205.379

(c) 5.8^3 is less than 200 and 5.9^3 is greater than 200, so the square root of 200 must be between 5.8 and 5.9.

(d) 5.85

3 (a) 6.32   (b) 2.71

Area

> **Essential**
>
> Angle measurer, set square, sheet 213
>
> **Practice booklet** pages 51 to 53

Ⓐ Area of a parallelogram (p 110)

The aim is for pupils to justify for themselves the formula for the area of a parallelogram and to gain confidence in using the correct dimensions. The work provides an opportunity for pupils to practise accurate drawing.

> Angle measurer, set square, sheet 213

◊ These points should emerge from dissecting the parallelogram.

• There are essentially two ways of dissecting the parallelogram, cutting at right angles to the longer side or at right angles to the shorter side.

• These two ways justify the two ways of measuring base and height for use in the formula.

The parallelogram has an area of about $64 \, \text{cm}^2$. It can be made into a rectangle 7 cm by 9.1 cm or 10 cm by 6.4 cm.

A set square should be used as a guide when measuring perpendicular to a side.

◊ The questions involving measuring and calculation can be used to make an important point about degree of accuracy. Pupils will record distances to the nearest 0.1 cm and, even then, there are likely to be variations across the class of 0.1 cm in the recorded values. The area calculated from the greatest recorded values will be noticeably different from the area calculated from the smallest recorded values. You can use this to demonstrate that the second place of decimals in an answer of this kind is meaningless, and the first decimal place is suspect; however, rounding to the nearest whole number may sometimes give too 'coarse' a result. (In the answers, some areas calculated from lengths given as decimals in the pupil's book have been left with two decimal places showing: this is to help you trace pupils' errors and does not represent an appropriate degree of accuracy when dealing with areas in practice.)

◊ The approach above breaks down in the case of an 'overhanging' parallelogram, which will make a rectangle if cut at right angles to its longer side but not its shorter side. However, the approach in question A3 shows that the 'base × height' formula still applies when the shorter side is the base.

B **Area of a triangle** (p 112)

Set square

◊ The main difficulty is identifying the correct lengths to multiply by. You could ask 'What parallelogram could the triangle be half of?'
There are of course three for every triangle:

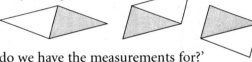

The next question is
'Which of the parallelograms do we have the measurements for?'

B2 Again, a set square should be used as a guide when measuring perpendicular to a 'base'.

C **Area of a trapezium** (p 114)

Set square

◊ The area of trapezium P can be found by adding areas and Q's area can be found by subtraction. However R's area cannot be found by either of these methods; pupils could work in pairs or small groups to find a way of working out its area and their conclusions could then be discussed with the whole class. Here are two approaches, though there are several others.

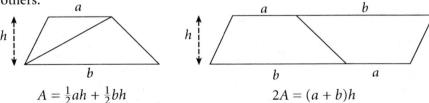

$$A = \tfrac{1}{2}ah + \tfrac{1}{2}bh \qquad\qquad 2A = (a + b)h$$

Some people find it easier to think of the formula for a trapezium as

Find the mean of the lengths of the parallel sides by adding them and dividing by 2, then multiply by the distance between them.

For this interpretation the formula can be written as $\quad A = \dfrac{(a + b)}{2}\, h$

Answers based on measurements can be expected to differ slightly from those given here.

Ⓐ **Area of a parallelogram** (p 110)

A1 About 75 cm²

A2 About 47 cm²

A3 P and R are equal in area, because A + P + B = R + A + B.

A4 (a) 10 cm² (b) 22.62 cm²
(c) 13.02 cm² (d) 29.11 cm²

A5

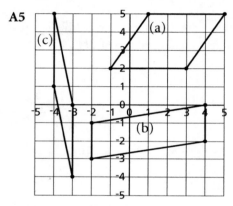

(a) 12 square units
(b) 12 square units
(c) 4 square units

A6 About 17.5 cm²

A7 $a = 3.0$ m $b = 2.5$ m $c = 2.5$ m
$d = 3.0$ m $e = 1.6$ m

*A8 $a = 2.7$ m $b = 5.0$ m

Ⓑ **Area of a triangle** (p 112)

B1 Parallelogram Triangle
(a) 27.3 cm² 13.65 cm²
(b) 23.2 cm² 11.6 cm²
(c) 20.4 cm² 10.2 cm²
(d) 6.44 cm² 3.22 cm²

B2 (a) Base = 5.0 cm and height = 6.0 cm or
Base = 6.1 cm and height = 4.9 cm or
Base = 7.0 cm and height = 4.2 cm
Area ≈ 15 cm²

(b) Base = 5.3 cm and height = 7.5 cm or
Base = 7.5 cm and height = 5.2 cm or
Base = 8.8 cm and height = 4.5 cm
Area ≈ 20 cm²

(c) Base = 4.9 cm and height = 7.2 cm or
Base = 7.6 cm and height = 4.7 cm or
Base = 10.3 cm and height = 3.5 cm
Area ≈ 18 cm²

B3 (a) 11.7 cm² (b) 10.64 cm²
(c) 10.03 cm² (d) 9.62 cm²

B4 6 cm

B5 $a = 2$ cm $b = 2.4$ cm

Ⓒ **Area of a trapezium** (p 114)

C1 (a) 20.0 cm² (b) 17.5 cm²
(c) 39.0 cm² (d) 21.7 cm²
(e) 15.3 cm²

C2 17.1 cm²

C3 (a) About 12.1 cm²
(b) About 23.9 cm²

What progress have you made? (p 116)

1 11.7 m²

2 6.0 m

3 About 6.6 cm²

4 42.75 m²

Practice booklet

Section A (p 51)

1 (a) 15 cm² (b) 7.0 cm²
(c) 22.4 cm² (d) 4.7 cm²
(e) 16.5 cm² (f) 17.5 cm²
(g) 25.5 cm² (h) 13.5 cm²

2 (a) 12 cm (b) 4.5 cm
(c) 4 cm (d) 6 cm

Section B (p 52)

1 12 cm^2: A B D E G L M
   18 cm^2: C H I J
   24 cm^2: F K N

2 (a) 11.76 cm^2 (b) 11.96 cm^2
   (c) 10.75 cm^2 (d) 11.22 cm^2
   (e) 14.28 cm^2 (f) 19.74 cm^2

Section C (p 53)

1 72 cm^2: A F G
   48 cm^2: B C
   60 cm^2: D E

Equivalent expressions

Optional
Sheet 192 (blank 'walls')

Practice booklet pages 54 to 59

𝔸 Walls (p 117)

> Optional: Sheet 192 (blank 'walls')

In these 'walls' the number on each 'brick' is found by adding the numbers on the two bricks below.

◊ Ask pupils to look at the first wall. Can they see how it has been constructed? What would the missing numbers be for all three walls?

A2, A3 Some pupils may be able to follow an algebraic argument to explain why the largest number should be placed in the middle brick on the bottom row.

The top brick contains two 'lots of b', so having b as the greatest of the three variables will maximise the total on the top brick.

	$a+2b+c$	
$a+b$		$b+c$
a	b	c

Pupils can investigate the general question of how to produce the largest total on the top brick with different sized walls. In one school, pupils investigated different walls with 1, 2, 3, 4 and 5 on the bottom row. They considered:
- How many different ways can the numbers be arranged?
- What is the range of totals for the top brick?
- Which arrangement gives the smallest/largest total?

A5, A6 These questions lead into the algebra in section B.

B Building in algebra (p 118)

Optional: Sheets 192 (blank 'walls')

◊ You could begin the discussion by showing pupils statements such as

- $n + 4 = 7$
- $n + n + n = 3n$
- $n + n + n = 6$
- $2n + 3n = 5n$
- $4n + 3 + n + 5 = 5n + 8$

and asking them to classify them as 'always true' or 'sometimes true'. Establish that, for example, $n + 4 = 7$ is only true for $n = 3$, but $2n + 3n = 5n$ is true for any value of n. Examples can be used to demonstrate that 2 times a number added to 3 times a number is equivalent to 5 times that number, such as $(2 \times 100) + (3 \times 100) = 5 \times 100$. Remind pupils that $2n + 3n$ and $5n$ are called **equivalent expressions** as they have the same value for *all* values of n.

Ask pupils to simplify where possible a set of expressions such as

- $x + x + x + x$
- $6x + 2x$
- $7x + 5x + 1 + 3$
- $2x + 7$

If pupils make mistakes, use numerical substitution to reveal them. For example, $2x + 7 \neq 9x$ because when $x = 2$, $2x + 7 = 11$ but $9x = 18$.

Remind pupils that order does not matter when adding ($a + 7 = 7 + a$).

◊ Show how we can use these techniques on the walls at the beginning of the section to give an expression for the number on the top brick. These walls give $n + 20$ and $6a + 51$ on their top bricks.

Ask pupils how this could help us solve problems such as

- For the left-hand wall, what is the number on the top brick if $n = 1$? You can refer them back to A6 (a) where they solved this problem with particular numbers and check the results are the same.
- For the left-hand wall, what value for n gives 100 on the top brick? Check this by completing the wall.
- For the right-hand wall, what is the number on the top brick if $a = 10$? Check this by completing the wall.

> 'Used this plan and it went well.'

ℂ **Perimeters** (p 120)

◊ Pupils often try to simplify expressions further where this is not possible. For example, a common mistake is $3x + 2y = 5xy$. Again, one way to show the approach is wrong is by substituting numbers.

◊ Remind pupils that in additions order does not matter, so $b + a + 2a + 4 + 3b = a + 2a + b + 3b + 4 = 3a + 4b + 4$.

𝔻 **Subtracting** (p 121)

◊ Remind pupils of work in *Book S1* 'Number grids': $n + 8 - 2 = n + 6$ because adding 8 and taking away 2 has the same effect as adding 6. Also $n - 2 - 8 = n - 10$ and $n - 8 + 2 = n - 6$.

'This is the hard part.'

◊ As for section B, you could show pupils statements such as
  • $3n - 2n = n$
  • $6 + 2n - 3n = 6 - n$
  • $6 - 3n + 2n = 6 - 5n$
  • $6 - 3n - 2n = 6 - 5n$
  • $8 - 2n = 6n$

and they classify them as 'always true', 'never true' or 'sometimes true'.

It may help to think about finding the value of each expression when $n = 1000$ or $n = 1\,000\,000$. For example, when $n = 1000$, $6 - 3n - 2n$ gives $6 - 3000 - 2000 = 6 - 5000$ and this may help pupils to see that $6 - 3n - 2n = 6 - 5n$.

◊ Pupils commonly think that, for example,
$6 - 3n + 2n = 6 - (3n + 2n) = 6 - 5n$

'Key points. These were important in teaching'.

Numerical examples can be used to show that subtracting three times a number and then adding two times that number is equivalent to subtracting that number, for example:
$5 - (3 \times 100) + (2 \times 100) = 5 - 100$.

◊ Emphasise that in calculations using addition and subtraction, order does not matter, but you must remember that '+' or '−' stays with the term being added or subtracted.

  For example, $7 + 5 - 1 - 3 = 7 - 3 + 5 - 1 = 7 - 1 - 3 + 5$

  Similarly $6 + 2x - 1 - 3x = 6 - 1 + 2x - 3x = 5 - x$

◊ Pupils can try to find pairs of equivalent expressions from the set on page 6 before discussing examples with more than one variable. Again, numerical examples may be helpful:
$7 - (5 \times 100) + (6 \times 24) + (2 \times 100) - (2 \times 24) = 7 - (3 \times 100) + (4 \times 24)$.

E Magic squares (p 121)

'Worked well. Some got muddled at points but this context allows self-checking. Worked well in pairs.'

Pupils consolidate work on simplifying linear expressions in the context of magic squares. Some work on simple substitution is included.

F Brackets (p 123)

Pupils review expanding brackets from *Book S1* 'Inputs and outputs' and then expand brackets of the form $a(bx + c)$ where b or c can be negative. The section ends with work on expanding brackets of the form $a(bx + cy)$ where a, b and c are all positive. You may wish to include examples like these in your introduction.

◊ To begin with, pupils could work in groups to sort the expressions at the start of the section into four pairs of equivalent expressions. They could justify their decisions to the rest of the class.

◊ A visual approach can help with positive coefficients.
For example, if a is the number of marbles in a bag, $2(2a + 3) = 4a + 6$ can be illustrated by the following diagram.

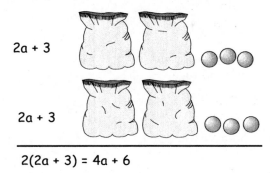

$$2a + 3$$

$$2a + 3$$

$$2(2a + 3) = 4a + 6$$

However, this approach is less useful when subtraction is involved.

◊ Pupils can think about calculations such as $6 \times (100 - 2)$ and try to explain how they know this is equivalent to $6 \times 100 - 6 \times 2$.
They can then apply this logic to algebraic expressions.

◊ An alternative approach (when the multiplier of the bracket is a whole number) is to use repeated addition.
For example $3(5 - 2n) = 5 - 2n + 5 - 2n + 5 - 2n = 15 - 6n$.

F9 For each shape, encourage pupils to include brackets in at least one expression.

18 Equivalent expressions • 79

Ⓐ Walls (p 117)

A1 (a)

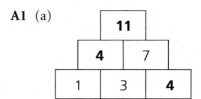

(b)

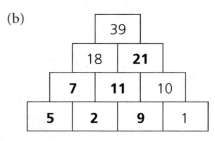

(c)

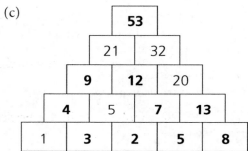

(b)

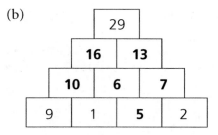

A2 (a) 18

(b) The pupil's wall and total

(c) Different possible totals are 23, 18 and 15.

A3 18 is the highest possible total.

A4 (a) 21

(b) The pupil's wall and total

(c) The lowest possible total is 17.

A5 (a)

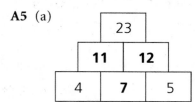

A6 (a) (i) 21

(ii) 24

(b)

Number on yellow brick	Number on top brick
1	21
2	22
3	23
4	24
5	25
10	30

(c) Add 20 to the number on the yellow brick to find the number on the top brick.

(d) 120

Ⓑ Building in algebra (p 118)

B1 (a)

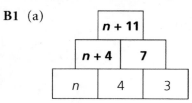

(b) (i) 16

(ii) The pupil's check

B2 (a)

(b) (i) $p = 5$

(ii) The pupil's check

B3 (a)

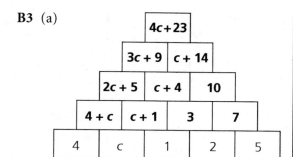

(b) 63

(c) (i) 3

　(ii) The pupil's check

B4 (a)

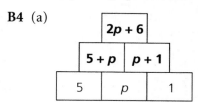

(b)

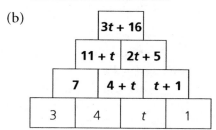

(c)

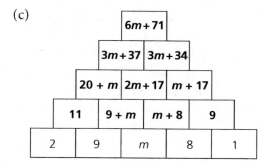

B5 (a)

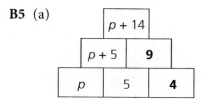

(b)

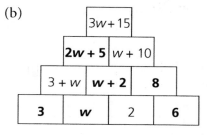

(c)

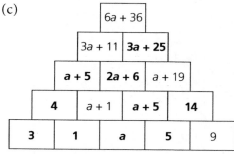

B6 (a) The pupil's walls
The numbers on the top bricks are the same.

(b) The pupil's walls
The numbers on the top bricks are the same again.

(c) The pupil's completed walls, both with $3n + 15$ on the top brick

ℂ Perimeters (p 120)

C1 (a) $3k$　　　　　(b) $3p + 2t$

(c) $2m + 5$　　　(d) $6c + d + 6$

(e) $2x + 8y + 11$

C2 (a) $2x + 3y$　　　(b) $5m + n$

(c) $8p + 5q$　　　(d) $8g + 8h$

(e) $7x + 5y + 7$　(f) $5p + 8q + 9$

C3 (a) Rectangle with lengths a, b, a, b

(b) Parallelogram with lengths
$m, 3k, m, 3k$

(c) Rectangle with lengths $4b, 3a, 4b, 3a$

(d) Square with lengths $2k, 2k, 2k, 2k$

(e) Kite with lengths $5, 3p, 3p, 5$

(f) Isosceles triangle with lengths
$2x, 7, 7$

(g) Kite with lengths $5z, 2y, 2y, 5z$

D Subtracting (p 121)

D1 (a) $4p$ (b) $5q$ (c) $r + 13$

     (d) $6t - 3$ (e) $2u$ (f) $3v - 1$

     (g) $3x$ (h) 7

D2 A and C, B and F, D and E

D3 (a) $2s + 3t$ (b) $4p + 9q$

     (c) $13m + n$ (d) $j + 3k$

     (e) $g + h - 3$ (f) $15e - 5f$

     (g) $3c + 3d$ (h) $a + 9b$

D4 (a) $10 - 2t$ (b) $7 - 8v$

     (c) $1 - w$ (d) $13 - 4x$

     (e) $5 - 7y$ (f) $3 - 8z$

D5 (a) $7m - 2n$ (b) $4r - 4s$

     (c) $10 - 8p + 3q$ (d) $3t - 4v - 3$

     (e) $w - x - 3$ (f) $8y - 4z$

E Magic squares (p 121)

E1 (a) Yes

     (b) No

E2 (a)

9	6	6
4	7	10
8	8	5

     (b)

11	5	7
7	8	15
7	9	7

     (c) Square (a) is magic.

E3 (a)

3	11	4
7	6	5
8	1	9

(b) Yes

(c) $w + 2 + 3w + 8 + 5 - w = 3w + 15$

(d) 2nd row:

$8 - w + 5 + w + 3w + 2 = 3w + 15$

3rd row:

$5 + 3w + 2 - w + w + 8 = 3w + 15$

(e) Each set of expressions adds to give $3w + 15$.

(f) Each row, column and diagonal adds to give the same expression so any value of w will give a magic square.

E4 (a)

4	21	14
23	13	3
12	5	22

(b) Yes

(c) $3a - 2b + 8a + 5b + 7a = 18a + 3b$

(d), (e) Each set of expressions adds to give $18a + 3b$.

(f) Each row, column and diagonal has the same total so any values of a and b will give a magic square.

E5 (a) This is not a magic square. For example, the total for the first column is $9x + 3$ but for the second column the total is $8x + 4$.

(b) This is not a magic square. For example, the total for the first column is $9y + 3z$ but the totals for the diagonals are $6z$ and $6y + 12z$.

(c) This is a magic square. The total each time is $15y - 6z$.

***E6** (a)

$8 + 6k$	$3 - 4k$	$4 + k$
$1 - 4k$	$5 + k$	$9 + 6k$
$6 + k$	$7 + 6k$	$2 - 4k$

(b)

2b	5a − 3b	4a − 2b
7a − 5b	3a − b	3b − a
2a	b + a	6a − 4b

F Brackets (p 123)

F1 (a) $5n + 15$ (b) $3m + 3$

     (c) $8 + 4p$ (d) $2q − 12$

     (e) $7r − 35$ (f) $60 + 10s$

     (g) $35 − 5t$ (h) $32 − 8v$

     (i) $9w − 18$

F2 A and I, B and D, C and F, E and G

F3 (a) $6q + 12$ (b) $10r + 2$

     (c) $8 + 20s$ (d) $15 + 20n$

     (e) $6m − 15$ (f) $12p − 20$

     (g) $14 − 4t$ (h) $12 − 32v$

     (i) $18w − 45$ (j) $3 − 6x$

     (k) $20 − 30y$ (l) $40z − 90$

F4 (a) $2(w + 5) = 2w + \mathbf{10}$

     (b) $4(\mathbf{2x} − 3) = 8x − 12$

     (c) $6(\mathbf{1} − y) = 6 − 6y$

     (d) $2(5 − \mathbf{3z}) = 10 − 6z$

F5 A and C

F6 Q and S

F7 A and F, B and H, C and D, E and G

F8 A and D

F9 Some possible answers are

     (a) $3(x + 2y)$, $3x + 6y$, $2y + 3x + 4y$

     (b) $2(2p + q)$, $4p + 2q$, $4p + q + q$

     (c) $4(2a + b)$, $2(4a + 2b)$, $8a + 4b$

What progress have you made? (p 125)

1 (a)

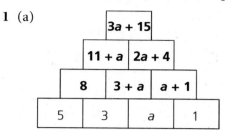

     (b) (i)   21   (ii) The pupil's check

     (c)   1

2 $5h + 11$

3 $8x + y + 10$

4 $5g + 5h + 5$

5 (a) $2v + 5$ (b) $6j + 7k$

     (c) $5 − 3w$ (d) $7y − z − 7$

6 (a) $30 − 5h$ (b) $6w + 15$

     (c) $12t − 4$ (d) $15 − 10b$

     (e) $6m + 6n$ (f) $6b + 2c$

Practice booklet

Section B (p 54)

1 (a)

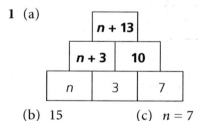

     (b)   15   (c) $n = 7$

2 (a)   Wall A

Wall B

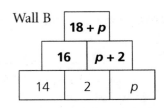

(b) (i) $p = 6$ top A = 22   top B = 24
    (ii) $p = 7$ top A = 24   top B = 25

3 (a)

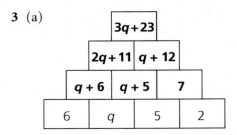

(b)

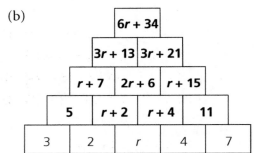

4 (a)

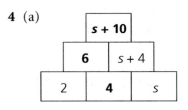

(b)

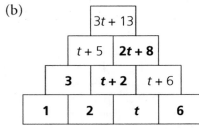

(c)

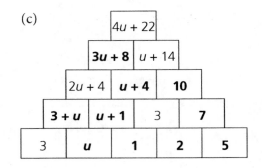

5 (a) $2a + 6$ (b) $3b + 3$ (c) $3x + 5$
   (d) $5y + 10$ (e) $7p + 10$ (f) $4j + 5$
   (g) $8g + 8$ (h) $4q + 11$

*6 Wall P

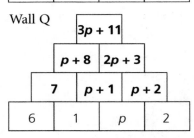

Wall Q

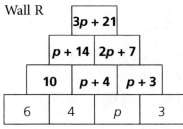

Wall R

Wall P and wall R will always give the same top number.

Section C (p 56)

1 (a) $20 + 4x + 4y$ (b) $4x + 4y$
   (c) $18 + 6p$ (d) $18y + 6z$

2 (a) $4g + 3h$ (b) $10p + 5q$

  (c) $6m + n + 5$ (d) $12 + 6t + u$

  (e) $5 + 9a + 4b$ (f) $7 + 3x + 6y$

3 (a)   (b)

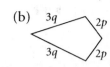

Section D (p 57)

1 (a) $2p + 9$ (b) $q + 4$ (c) $11r - 2$

  (d) $7s + 3$ (e) $t + 4$ (f) $4u + 2$

2 A and F $(3p + 2)$,   B and E $(2p + 1)$,
  C and D $(3p + 4)$

3 (a) $2v + 8w$ (b) $2 + 2p$ (c) $5 + 2a$

  (d) $5q - r$ (e) $4t - 2u$ (f) $10 - 2y$

4 C simplifies to $a - b$.

5 (a) $6p - 5q$ (b) $8 - 5r + 3s$

  (c) $10 - t + 3u$ (d) $2v - 11w$

  (e) $8a - 11$ (f) $p + q$

  (g) x (h) $7g$

*6 $8y - 3x$

Section E (p 58)

1 (a) All rows, columns, diagonals sum to
    $15p$.

  (b)

17	38	5
8	20	32
35	2	23

  (c) $p = 2$

2 Square A $(9a)$

3 (a) $48 - 3k$

  (b) (i) $k = 2$

15	0	27
26	14	2
1	28	13

     (ii) $k = 1$

16	6	23
22	15	8
7	24	14

Section F (p 59)

1 (a) $4p + 20$ (b) $3q + 6$ (c) $18 + 6r$

  (d) $8s - 16$ (e) $15 - 5t$ (f) $12u - 48$

2 (a) $4x + 8$ (b) $12 + 9y$ (c) $30 + 15z$

  (d) $27a - 36$ (e) $10b - 15$ (f) $33 - 22c$

3 (a) $4(d + 3) = 4d +$ **12**

  (b) $3(5 - 2e) = 15 -$ **6e**

  (c) $8(\mathbf{f} - 2) = 8f - 16$

  (d) $4(\mathbf{g} - 9) = 4g - 36$

4 C and D

5 A and E, B and C, D and F.

6 Two expressions from
  $4(p + 2q)$, $4p + 8q$, $2(2p + 4q)$...

No chance!

This unit revises finding probabilities for single events and progresses to multiple events where lists and grids are used to record all the outcomes.

Essential	**Optional**
Tetrahedral dice (two per pair)	A set of month name cards
Counters (two per pair, different colours)	Sheet 214 (to make tetrahedral dice)
Sheet 206	
Practice booklet pages 60 to 62	

🅐 **Probabilities** (p 126)

> Optional:
> A set of month name cards

T

◊ The first page is intended as a starter to remind pupils of earlier work on probability.

The main points to establish are that all the mutually exclusive probabilities sum to 1 and therefore the probability that something does not happen is 1 minus the probability that it does.

◊ Prime, square and triangle numbers are referred to in this section and may need revising.

🅑 **All the outcomes** (p 128)

T

◊ Thinking about these games leads to a need for systematically listing the equally likely outcomes. You may decide to move straight to using a grid; however, some of the questions in section B involve three throws, where a grid is not feasible. Grids are introduced in section C.

◊ You could play one or more of these games with the class, but the results from a short run of games would be inconclusive. In each case all the equally likely outcomes need to be listed.

Coin game

The teacher is being unfair. Of the four possible outcomes HH, HT, TH, TT, the teacher wins on two and the pupil on one; the fourth outcome leads to another round where the teacher again has the advantage.

Beat that!

This game is fair. Omitting the 6 from the teacher's throw, there are 30 equally likely outcomes of the pair of throws. Half of these lead to a win for the pupil.

Dice difference

This game is unfair. Of the 36 equally likely outcomes, 16 give a difference of 0 or 1.

B6 In reality the probability of a male birth is about 0.515.

C Listing with grids (p 130)

> Tetrahedral dice (two per pair): you can make them from sheet 214, or you can use ordinary dice and ignore throws of 5 or 6
> Counters (two per pair, different colours)
> Sheet 206 (or pupils can easily draw the board)

Pursuit

◊ Pupils should play the game a few times in order to see what decisions need to be made as they play.

◊ Players need to know the relative likelihood of the different total scores with the two four-sided dice. Introduce a grid for showing all the possible outcomes and their totals, as at the foot of page 130.

Pupils should now be able to decide on the best move in the games A to D shown on the page.

D Which is more likely? (p 132)

◊ You may need to check that pupils are confident in converting a fraction to decimals..

A Probabilities (p 126)

A1 (a) $\frac{1}{12}$ (b) $\frac{11}{12}$

A2 (a) $\frac{1}{2}$ (b) $\frac{1}{2}$ (c) $\frac{5}{12}$
     (d) $\frac{7}{12}$ (e) $\frac{1}{3}$ or $\frac{4}{12}$ (f) $\frac{2}{3}$ or $\frac{8}{12}$

A3 (a),(b) The answers sum to 1.
          One or other must happen so the
          total must be 1.

A4 (a) 1,4,9   (b) $\frac{1}{4}$ or $\frac{3}{12}$ (c) $\frac{3}{4}$ or $\frac{9}{12}$

A5 (a) 1,3,6,10   (b) $\frac{1}{3}$ or $\frac{4}{12}$ (c) $\frac{2}{3}$ or $\frac{8}{12}$

A6

	Happening	Not happening
(a)	$\frac{25}{49}$	$\frac{24}{49}$
(b)	$\frac{1}{7}$ or $\frac{7}{49}$	$\frac{6}{7}$ or $\frac{42}{49}$
(c)	$\frac{15}{49}$	$\frac{34}{49}$
(d)	$\frac{9}{49}$	$\frac{40}{49}$
(e)	$\frac{9}{49}$	$\frac{40}{49}$

A7 (a) $\frac{7}{15}$ (b) $\frac{1}{3}$ or $\frac{5}{15}$ (c) 0
     (d) $\frac{4}{5}$ or $\frac{12}{15}$ (e) $\frac{8}{15}$ (f) $\frac{1}{5}$ or $\frac{3}{15}$

A8 $\frac{2}{5}$

A9 $\frac{4}{25}$

A10 16

A11 49

B All the outcomes (p 128)

B1 (a) The pupil's list of 9 outcomes
     (b) (i) $\frac{1}{9}$ (ii) $\frac{1}{3}$ or $\frac{3}{9}$ (iii) $\frac{1}{3}$ or $\frac{3}{9}$

B2 (a) The pupil's list of 8 outcomes
     (b) (i) $\frac{1}{8}$ (ii) $\frac{3}{8}$
        (iii) $\frac{1}{4}$ or $\frac{2}{8}$ (iv) $\frac{1}{2}$ or $\frac{4}{8}$ (v) 0

B3 (a) The pupil's list of 12 outcomes
     (b) $\frac{1}{12}$ (b) $\frac{1}{3}$ or $\frac{4}{12}$

B4 (a) The pupil's list of 8 outcomes
     (b) $\frac{1}{8}$ (c) $\frac{1}{4}$ or $\frac{2}{8}$

B5 (a) The pupil's list of 27 outcomes
     (b) (i) $\frac{1}{27}$ (ii) $\frac{1}{9}$ or $\frac{3}{27}$
        (iii) $\frac{2}{3}$ or $\frac{18}{27}$ (iv) $\frac{2}{9}$ or $\frac{6}{27}$

B6 (a) $\frac{1}{4}$ (b) $\frac{1}{2}$

C Listing with grids (p 130)

C1 (a) 0   (b) $\frac{1}{8}$ or $\frac{2}{16}$
     (c) $\frac{13}{16}$ (d) $\frac{1}{16}$

C2 The pupil's grid
     (a) $\frac{1}{6}$ or $\frac{6}{36}$ (b) $\frac{1}{12}$ or $\frac{3}{36}$ (c) $\frac{1}{36}$
     (d) $\frac{1}{2}$ or $\frac{18}{36}$ (e) $\frac{15}{36}$ (f) $\frac{1}{12}$ or $\frac{3}{36}$

C3 (a) 7   (b) 2 and 12
     (c)

Score	2	3	4	5	6	7	8	9	10	11	12
Prob	$\frac{1}{36}$	$\frac{2}{36}$	$\frac{3}{36}$	$\frac{4}{36}$	$\frac{5}{36}$	$\frac{6}{36}$	$\frac{5}{36}$	$\frac{4}{36}$	$\frac{3}{36}$	$\frac{2}{36}$	$\frac{1}{36}$

C4 (a) The pupil's grid
     (b) (i) $\frac{2}{9}$ or $\frac{8}{36}$ (ii) $\frac{5}{18}$ or $\frac{10}{36}$
        (iii) $\frac{1}{6}$ or $\frac{6}{36}$ (iv) $\frac{1}{6}$ or $\frac{6}{36}$ (v) 0

C5 The pupil's grid
     (a) $\frac{1}{5}$ or $\frac{5}{25}$ (b) $\frac{4}{25}$ (c) $\frac{9}{25}$
     (d) $\frac{12}{25}$ (e) $\frac{1}{25}$ (f) $\frac{3}{5}$ or $\frac{15}{25}$

D Which is more likely? (p 132)

D1 The probabilities are
     A 0.267 (3 d.p.)   B 0.292 (3 d.p.)   C 0.3
     So C has best chance of winning.

D2 The probabilities are
     A 0.333 (3 d.p.)   B 0.286 (3 d.p.)
     So A is more likely.

D3 The probabilities are
     A 0.3125   B 0.255 (3 d.p.)
     So A is more likely.

D4 The probabilities are
     A 0.722 (3 d.p.)   B 0.618 (3 d.p.)
     So A is more likely.

What progress have you made? (p 133)

1 (a) $\frac{1}{15}$ (b) $\frac{7}{15}$

     (c) $\frac{1}{3}$ or $\frac{5}{15}$ (d) $\frac{1}{3}$ or $\frac{5}{15}$

2 (a) The pupil's list of 27 outcomes

     (b) (i) $\frac{1}{27}$ (ii) $\frac{1}{9}$ or $\frac{3}{27}$

         (iii) $\frac{2}{9}$ or $\frac{6}{27}$ (iv) $\frac{7}{9}$ or $\frac{21}{27}$

3 (a)

	1	2	3	4	5
1	●				
2		●			□
3			●	□	
4			□	●	
5		□			●

     ● = show same number

     □ = total is 7

     (b) (i) $\frac{1}{5}$ or $\frac{5}{25}$ (ii) $\frac{4}{25}$

Practice booklet

Section A (p 60)

1 Red

2 (a) $\frac{1}{10}$ (b) $\frac{9}{10}$

3 (a) $\frac{2}{10} = \frac{1}{5}$ (b) $\frac{5}{12} = \frac{1}{2}$ (c) $\frac{5}{12} = \frac{1}{2}$

     (d) 0   (e) $\frac{8}{10} = \frac{4}{5}$

4 (a) $\frac{7}{20}$ (b) $\frac{13}{20}$ (c) $\frac{5}{20} = \frac{1}{4}$

     (d) 0   (e) $\frac{12}{20} = \frac{6}{10} = \frac{3}{5}$ (f) $\frac{13}{20}$

5 (a) 12   (b) 8   (c) $\frac{2}{3}$

     (d) $\frac{4}{24} = \frac{1}{6}$ (e) $\frac{5}{6}$

Section B (p 61)

1 (a) The pupil's list of 16 outcomes

     (b) (i) $\frac{1}{4}$ (ii) $\frac{3}{4}$ (iii) $\frac{9}{16}$

2 (a)

     14   16   19

     41   46   49

     61   64   69

     91   94   96

     (b) 12 different numbers can be made.

     (c) (i) $\frac{6}{12} = \frac{1}{2}$ (ii) $\frac{3}{12} = \frac{1}{4}$ (iii) $\frac{3}{12} = \frac{1}{4}$

        (iv) $\frac{3}{12} = \frac{1}{4}$ (v) $\frac{2}{12} = \frac{1}{6}$ (vi) $\frac{9}{12} = \frac{3}{4}$

Sections C and D (p 62)

1

Square spinner	Pentagon spinner				
	1	**2**	**3**	**4**	**5**
1	2	3	4	5	6
2	3	4	5	6	7
3	4	5	6	7	8
4	5	6	7	8	9

2 (a) $\frac{3}{20}$ (b) $\frac{1}{20}$

     (c) $\frac{3}{20}$ (d) $\frac{4}{20} = \frac{1}{5}$

3 5 and 6 are the most likely totals.

4 (a)

Triangle spinner	Hexagon spinner					
	1	**2**	**3**	**4**	**5**	**6**
1	1	2	3	4	5	6
2	2	4	6	8	10	12
3	3	6	9	12	15	18

     (b) 6 is the most likely score.

5 The probabilities are

     A   0.33 (to 2 d.p.)   B   0.43 (to 2 d.p.)

     So B is more likely

20 Recipes

Pupils use unitary methods to calculate quantities.

Practice booklet pages 63 and 64

A How much? (p 134)

The problems on page 134 are intended to be tackled without any help or pre-teaching. It is hoped that after thought and discussion in groups, pupils will arrive at the unitary method for themselves.

It is not intended that pupils get bogged down in the calculations so calculators can be used throughout.

◊ One way to structure the initial session is:
  • pupils think about the problems individually and possibly make notes;
  • the problems are then discussed in small groups;
  • the methods used to solve the problems are then discussed and evaluated by all the pupils in a whole-class discussion.

Leek and potato soup

For 12 people	For 3 people
60 g butter	15 g butter
800 g potatoes	200 g potatoes
1560 ml veg stock	390 ml veg stock
180 g cheese	45 g cheese
6 leeks	$1\frac{1}{2}$ leeks

Blackberry and apple crumble

For 2 people	For 9 people
100 g blackberries	450 g blackberries
160 g cooking apples	720 g cooking apples
20 g sugar	90 g sugar
60 g flour	270 g flour
40 g margarine	180 g margarine

Semolina-Halva

For 6 people	For 3 people
210 g butter	105 g butter
300 g semolina	150 g semolina
180 g pine kernels	90 g pine kernels
480 g sugar	240 g sugar
900 ml milk	450 ml milk

◊ The problems on page 135 could also be discussed by the pupils working in small groups or pairs. Pupils need to know the number of people in their class and in the whole school, and the number of teachers in the school, for the last three questions.

B **The unitary method** (p 136)

The unitary method is set out formally. A calculator is essential for some problems.

◊ The panel above B6 shows that it is important not to round in the middle of a calculation. You could ask pupils to work out what the result is when the weight of one sweet is rounded to 8.5 g.

A **How much?** (p 134)

A1 40 g

A2 1200 g

A3 300 g

A4 45 g

A5 800 g

A6 4 leeks

A7 80 g

A8 100 g

A9 150 g

A10 400 g

A11 880 g

A12 300 ml

A13 75 g

A14 90 g

A15 560 g

A16 6110 ml

A17 1875 g

A18 1300 g

A19 1320 g

A20 The pupil's answer for the class

A21 The pupil's answer for the school

A22 The pupil's answer for the teachers in the school

B **The unitary method** (p 136)

B1 (a) 30 g   (b) 270 g
(c) 1200 ml   (d) 400 g

B2 (a) 8 kg   (b) 40 kg

B3 X: 75 cm, 10 kg; Z: 165 cm, 22 kg

B4 200 cm

B5 (a) Boric acid 42 g, zinc sulphate 1.2 g
(b) Boric acid 112 g, zinc sulphate 3.2 g

B6 317 g

B7 817 g

B8 30.4 cm

B9 2813 g pumpkin (2812.5 g)
313 g butter (312.5 g)
1406 g potatoes (1406.25 g)

What progress have you made? (p 138)

1 (a) 210 ml (b) 350 ml

2 £3.85

3 83 cm

Practice booklet

Sections A and B (p 63)

1 200 g

2 1575 g

3 375 ml

4 175 g

5 750 g

6 180 ml

7 600 g

8 750 g

9 176 g plain flour
320 ml milk
120 ml water
48 ml melted butter

10 275 g plain flour
500 ml milk
187.5 ml water
75 ml melted butter

11 (a) 12 (b) 300

12 (a) 13.75 g (b) 4125 g

13 132 g

14 19 kg

*15 3242

21 Substituting into formulas

Essential

Sheet 215 (probably best enlarged on to A3 paper)
Dice marked 0, 0, 2, 2, 4, 4 and 3, 3, 5, 5, 7, 7 (one each per group)
Counters (one per pupil)

Practice booklet pages 65 to 68

Ⓐ **Review** (p 139)

Ⓑ **The right order** (p 140)

◊ One way of introducing this section is to let pupils work out the value of some expressions unaided, and then to discuss the answers they have come up with. Include some expressions with negative values. You can then point out the pitfalls that can arise.

It will be helpful to talk about problems that may come up when using a calculator to evaluate expressions.

Ⓒ **Including squares** (p 142)

Sheet 215 (better enlarged on to A3 paper)
Dice marked 0, 0, 2, 2, 4, 4 and 3, 3, 5, 5, 7, 7 (one each per group)
Counters (one per pupil)

◊ You may need to sort out the common misconception that, say, $4a^2$ can be evaluated by multiplying a by 4 and then squaring.

$\mathbb{D}$ Forming formulas (p 144)

$\mathbb{E}$ Words to symbols (p 146)

When introducing this section, emphasise once more that the letters always stand for numbers, and are not shorthand for names of objects. In question E3, for example, where the number of pancakes that Kevin has is $4p$, you may wish to draw attention to the fact the $4p$ does not stand for four pancakes.

$\mathbb{A}$ Review (p 139)

A1 (a) £290

   (b) (i) 100 square metres (m²)

      (ii) £650

   (c) £1184

A2 (a) (i) 390   (ii) £115   (iii) £225

   (b) (i) $s = 590$; $w = 975$; $m = 650$;
      $t = 20$; $d = 165$

     (ii) $s = 4990$; $w = 7575$; $m = 5050$;
      $t = 240$; $d = 1265$

A3 (a) 49 (b) 14 (c) 8 (d) 45
   (e) 3   (f) 39 (g) 13 (h) 2

A4 (a) 64 (b) 65 (c) 60 (d) 24

$\mathbb{B}$ The right order (p 140)

B1 (a) 16 (b) 4   (c) 16 (d) ⁻4

B2 (a) 28 (b) 36 (c) 48 (d) 0

B3 (a) 6   (b) 0   (c) 9   (d) 10

B4 (a) 16 with the pupil's working
   (b) 10 with the pupil's working

B5 (a) 4   (b) 2   (c) 6   (d) 5

B6 (a) 2   (b) 6   (c) 6   (d) 2

B7 (a) 50   (b) 60 cm   (c) 69 cm
   (d) 100 cm (e) 14 tonnes

B8 (a) 95   (b) 98°C   (c) 97°C
   (d) 96°C   (e) 70°C

$\mathbb{C}$ Including squares (p 142)

C1 (a) 25 (b) 27 (c) 50 (d) 60
   (e) 75 (f) 79 (g) 25 (h) 80

C2 (a) 50 (b) 20 (c) 14 (d) 2

C3 (a) 1 with the pupil's working
   (b) 120 with the pupil's working

C4 $z = 3y + 6$, $z = 3y^2$, $z = 18 - 3y$,
   $z = 6(4 - y)$

C5 $a = 2(b - 20)$, $a = 10 - 2b$, $a = 100 - 2b^2$

C6 16

C7 (a) 100 (b) 16   (c) 121 (d) 81

C8 (a) 2   (b) 15 (c) 17 (d) 32
   (e) 2   (f) 8   (g) 25 (h) ⁻3
   (i) 38 (j) 2   (k) 2   (l) 9

C9 (a) 12   (b) 3 m   (c) 27 m
   (d) 65 m (e) Yes   (f) 60 mph

$\mathbb{D}$ Forming formulas (p 144)

D1 (a) 22

   (b) The pupil's completed table with
      7, 10, **13**, **16**, 19, **22** in the second
      column (number of sandwiches)

   (c) $s = 3p + 4$

D2 (a) 19   (b) 9

(c) The pupil's completed table with **7, 9, 11, 13, 15, 17, 19** in the second column (number of new tyres)

(d) $t = 2b + 5$

D3 (a) £60   (b) £75

(c) $c = 6w + 30$ or equivalent

(d) £132   (e) 12 tonnes

D4 (a) 25   (b) 100

(c) Pattern 8   (d) $d = n^2$

(e) 225   (f) Pattern 13

D5 (a) 27   (b) $d = n^2 + 2$

(c) 402   (d) Pattern 18

E Words to symbols (p 146)

E1 (a) $2n$ (b) $n + 2$

E2 (a) $3n$ (b) $n - 3$ (c) $3n - 5$

E3 (a) $4p$ (b) $2p + 12$

E4 $25 - 4c$

E5 $3(10 - r)$

E6 (a) $3n$ (b) $n - 20$

E7 (a) 90   (b) $100 - z$ (c) $100 - 2z$

What progress have you made (p 148)

1 (a) 6   (b) 24   (c) 3

2 (a) 5   (b) 3   (c) 8

3 (a) 10   (b) 26   (c) 32

4 (a) 17   (b) 48   (c) 14

5 70

6 (a) 75 cm   (b) $w = h + 15$

7 (a) $3s$ (b) $s - 50$

Practice booklet

Section B (p 65)

1 (a) 12   (b) 20   (c) 24   (d) 2

(e) 4   (f) 8   (g) 0   (h) 4

2 (a) 10   (b) 24   (c) 2   (d) 5

(e) 13   (f) 12

3 (a) 26   (b) 10

(c) 18 cm   (d) $7\frac{1}{2}$ hours

4 (a) 15   (b) 10 minutes

(c) 20 minutes   (d) 24 square metres

Section C (p 66)

1 (a) 9   (b) 15   (c) 27   (d) 8

(e) 16   (f) 6   (g) 3   (h) 5

(i) 1   (j) 1   (k) 49   (l) 0

2 (a) 128   (b) 16   (c) 3   (d) 2

(e) 13   (f) 8

3 (a) $4(p + 3)$ is the highest; 28

(b) $20 - p^2$ is the highest; 19

(c) $4p - 10$ is the lowest; $^-10$

4 $y = 3 - 2x, y = 3(2 - x)$

Sections D and E (p 67)

1 (a) 13

(b) The pupil's completed table with **5, 7, 9, 11, 13, 15** in the second column (number of cereal bars)

(c) $c = 2p + 3$

2 (a) 18

(b) The pupil's completed table with **12, 15, 18, 21, 24, 27, 30** in the second column (number of chocolates)

(c) $c = 3p + 6$

3 (a) 9 cm^2 (b) $A = n^2$

4 (a) $n + 3$ (b) $3n$ (c) $n - 5$

5 (a) $100 - n$ (b) $100 - 2n$

Review 3 (p 149)

1 (a) 49 (b) 125 (c) 9 (d) ⁻8

2 (a) $\frac{1}{8}$ (b) $\frac{1}{8}$ (c) $\frac{3}{8}$ (d) $\frac{2}{8}$ or $\frac{1}{4}$
(e) $\frac{6}{8}$ or $\frac{3}{4}$ (f) 0

3 (a) 400 (b) 1 000 000

4 (a) $13a + 6$ (b) $5b + 7c + 2$
(c) $6d + 2e$ (d) $5f - g + 2$
(e) $4h + 12$ (f) $10j - 15$

5 $5 \times 2 = 10\,\text{cm}^2$

6 (a) H1 H2 H3 H4 H5
T1 T2 T3 T4 T5
(b) $\frac{1}{10}$
(c) $\frac{3}{10}$

7 $28\,\text{cm}^2$

8 (a) 11 (b) 35 (c) 3 (d) 18
(e) 19 (f) 25 (g) 22 (h) 8

9 55 cm

10 $h = 8\,\text{cm}$

11 (a) 9, ⁻9 (b) 4 (c) ⁻1

12 (a)

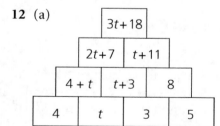

(b) 27 (c) 8

13 (a) $b + 10$ (b) $2b$
(c) $2(b + 10)$ or $2b + 20$

14 (a) $4x + 20$ or $4(x + 5)$ cm
(b) 44 cm (c) 10

15 (a) 60
(b) (i) 0.15 or $\frac{15}{100}$
(ii) 36

16 $22.68\,\text{cm}^2$

17

First spin \ Second spin	1	2	5	10	20
1	2	3	6	11	21
2	3	4	7	12	22
5	6	7	10	15	25
10	11	12	15	20	30
20	21	22	25	30	40

(a) $\frac{2}{25}$ (b) $\frac{12}{25}$ (c) $\frac{3}{25}$
(d) $\frac{8}{25}$ (e) $\frac{2}{25}$ (f) 12%

18 (a) 100 g (b) 125 g (c) 730 g
(d) 2:7

19 45 m

20 (a) $\frac{5}{8}$ (b) 40

Mixed questions 3 (Practice booklet p 69)

1 $36 + 64 = 100$

2 (a) $a + 2b$ (b) $4x - 3y + 4$
(c) $t - 17$

3 (a) $4r + 8$ (b) 13

4 (a)

1	5	6	11	17	28

(b) (i)

n	1	$n + 1$	$n + 2$	$2n + 3$	$3n + 5$

(ii) 11
(iii) $n = 4$

5 (a) $3n$ (b) $3n - 2$
(c) $2(3n - 2)$ or $6n - 4$

6 (a) $4x + 12$ (b) $10r - 15t$
(c) $18a - 42$

7 (a) $29.45\,\text{cm}^2$ (b) $21\,\text{cm}^2$

8 (a) 114 (b) 106 or 107

9 $(⁻10)^3$, $(⁻2)^3$, 1^2, $(⁻2)^2$, $(⁻7)^2$, 10^2, 5^3
or ⁻1000, ⁻8, 1, 4, 49, 100, 125

10 (a) 125°

  (b) The pupil's accurate drawing of parallelogram

  (c) Approx 17.2 cm²
    Either 3 × 5.7 = 17.1
    or 7 × 2.5 = 17.5

11 (a) 18   (b) 3   (c) 50   (d) ⁻8
  (e) 4   (f) 11   (g) 6   (h) 81

12 $p = 3n + 4$

13 (a) 180 g   (b) 262.5 g
  (c) 750 ml   (d) 3 : 7

14 (a) 1 line $\frac{2}{40} = 0.05$

    2 lines $\frac{14}{40} = 0.35$

    3 lines $\frac{21}{40} = 0.525$

    4 lines $\frac{3}{40} = 0.075$

    with the pupil's check
    (0.05 + 0.35 + 0.525 + 0.075 = 1)

  (b) 87.5; approximately 87 or 88 times

15 A $\frac{3}{8} = 0.375$ B $\frac{15}{36} = 0.417$
    B is more likely.

Scaling

Essential

Centimetre squared paper, other squared paper
Pieces of string about 45 cm long
Sheets 216 and 197

Practice booklet pages 72 to 76

A Spotting enlargements (p 152)

> Optional: squared paper

◊ With the photos of Bert, explain to pupils that they have to sort out the real enlargements from the fakes. They are not expected to make any measurements of the photos. Encourage them to say why they think particular photos are or are not enlargements of the original. They may use reference points in the fish photos (the fish's tail is above his elbow/above the bottom of Bert's jacket/below the bottom of his jacket); some may use proportions (the fish is less than half Bert's height) and this is to be encouraged.

B, D and G are enlargements; the rest are not.

A1 Pupils can work in small groups for this question. Encourage them to say why they think particular shapes are or are not enlargements of the one at the top of the page. A common error is to regard D, for example, as an enlargement of the original because the vertical and horizontal bars have been lengthened by one square.

◊ Some schools have used the following activity with pupils in small groups, working on squared paper. Ask each group to draw a simple shape such as one of these and label it 'original'.

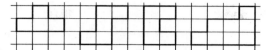

The group then draw three different enlargements of their shape. They also draw two shapes that look as though they could be enlargements of it but are not quite (the aim being to catch someone out). When the group are happy with what they have drawn they display their shapes mixed up on a single sheet of paper and show it to another group, who must say which are enlargements of the original and which are not.

B **Enlarging a shape** (p 155)

> Centimetre squared paper, sheets 216 and 197

◊ This provides practice in drawing to a given scale factor. The ideas that all lengths will be multiplied by the scale factor and that angles stay the same should emerge.

B3 The 'enlarging logos' activity has proved to be an attractive piece of consolidation work in which pupils can use their own creativity.
If pupils have no ideas for a logo they could draw a stylized version of their initials on squared paper.

C **Scaling down as well as up** (p 156)

> Squared paper

The term 'scaling down' is used here because many pupils find the idea of a 'half times enlargement' confusing, while a 'half times reduction' falsely implies that that a different mathematical process is being employed when the scale factor is less than 1.

◊ You can ask pupils to identify enlargements or scalings among the green shapes (for example, 'G is an enlargement of B with a scale factor 2'). You can leave the search open-ended or say, for example, 'Find me an enlargement with scale factor $\frac{1}{3}$ ... Can you find more than one?' or 'What scale factor enlarges D to F?'

Whenever pupils give an answer, encourage them to say how they got it.

Enlargements by, for example, $1\frac{1}{2}$ and $2\frac{1}{2}$ are relatively easy to spot and understand, but it may be useful to bring in the idea that $1\frac{1}{2}$ can be rewritten as the improper fraction $\frac{3}{2}$, giving a clue that this is to do with finding one-half of a length and then multiplying by 3.

If pupils seem confident you can move on to scale factors like $\frac{3}{4}$ and $\frac{2}{3}$.

The complete set of scale factors is as follows.

		To						
	A	B	C	D	E	F	G	H
A		$\frac{5}{8}$	$\frac{3}{8}$	$\frac{1}{2}$	$\frac{1}{4}$	$\frac{3}{4}$	$\frac{5}{4}$	$\frac{3}{2}$
B	$\frac{8}{5}$		$\frac{3}{5}$	$\frac{4}{5}$	$\frac{2}{5}$	$\frac{6}{5}$	2	$\frac{12}{5}$
C	$\frac{8}{3}$	$\frac{5}{3}$		$\frac{4}{3}$	$\frac{2}{3}$	2	$\frac{10}{3}$	4
From **D**	2	$\frac{5}{4}$	$\frac{3}{4}$		$\frac{1}{2}$	$\frac{3}{2}$	$\frac{5}{2}$	3
E	4	$\frac{5}{2}$	$\frac{3}{2}$	2		3	5	6
F	$\frac{4}{3}$	$\frac{5}{6}$	$\frac{1}{2}$	$\frac{2}{3}$	$\frac{1}{3}$		$\frac{5}{3}$	2
G	$\frac{4}{5}$	$\frac{1}{2}$	$\frac{3}{10}$	$\frac{2}{5}$	$\frac{1}{5}$	$\frac{3}{5}$		$\frac{6}{5}$
H	$\frac{2}{3}$	$\frac{5}{12}$	$\frac{1}{4}$	$\frac{1}{3}$	$\frac{1}{6}$	$\frac{1}{2}$	$\frac{5}{6}$	

D Scales for maps and drawings (p 158)

Pieces of string about 45 cm long

D7 Some pupils may be unfamiliar with the term 'as the crow flies'.

D8 Some may be content for their boats to travel across land.

◊ If pupils have had little experience of making their own scale drawings it would be worth them doing some of simple classroom objects.

◊ Questions like D18 and D19, where the scale has to be identified, pose something of a challenge and pupils can use various valid ways of reasoning. You could ask pupils to work on these in small groups and then have representatives from a few of the groups explain their reasoning.

E Ratios (p 163)

◊ Emphasise that ratios involve pure numbers, so that when the quantities to be compared are given in different units, they must first be expressed in the same units.

A Spotting enlargements (p 152)

A1 B, E and H are enlargements;
the rest are not.

B Enlarging a shape (p 155)

B1 (a) The pupil's enlargement
(b) They are equal.

B2 The pupil's enlargements and checks

B3 The pupil's logo and enlargement

B4 (a) The pupil's copy and enlargement
(b) About 10.5 cm
(c) About 21.0 cm, twice the perimeter
of the original triangle

B5 $a = 1$ cm, $b = 9$ cm, $c = 12$ cm,
$d = 6$ cm, $e = 4$ cm, $f = 2$ cm

B6 The pupil's enlargement on sheet 216.

B7 The pupil's enlargement on sheet 197.

C Scaling down as well as up (p 156)

C1 The pupil's drawing

C2 The pupil's drawing

C3 The pupil's drawing

C4 D

C5 F

C6 (a) $\frac{1}{5}$ (b) $\frac{1}{2}$ (c) $\frac{1}{4}$
(d) $\frac{1}{3}$ (e) $\frac{1}{4}$

*C7 (a) $1\frac{1}{2}$ or $\frac{3}{2}$ (b) $\frac{2}{3}$ (c) $\frac{3}{4}$
(d) $\frac{4}{5}$ (e) $2\frac{1}{2}$ or $\frac{5}{2}$

D Scales for maps and drawings (p 158)

D1 (a) 200 m   (b) 300 m
(c) 500 m   (d) About 125 m
(e) About 275 m

D2 (a) About 2100 m (b) About 2.1 km

D3 50 metres

D4 10 metres

D5 (a) About 33 m   (b) About 46 m
(c) About 33 m

D6 (a) The pupil's scale drawing
(b) 4.5 cm by 1.5 cm

D7 (a) 80 miles   (b) 128 miles
(c) 42 miles   (d) 32 miles
(e) 142 miles   (f) 104 miles
(g) 128 miles   (h) 134 miles

D8 (a) About 240 miles
(b) About 200 miles

D9 Birmingham, Portsmouth

D10 Empire State Building 380 m
Chrysler Building   320 m
Canada Tower   245 m
Sears Tower   440 m
Petronas Towers   450 m
(These heights include the spires.)

D11 (a) The pupil's scale drawing
(b) About 14.4 cm
(c) About 7.2 m

D12 (a) The pupil's scale drawing
(b) About 13.5 cm
(c) About 2.7 m

D13 (a) 4 cm   (b) 6.5 cm

D14 (a) 5 cm   (b) 7.6 cm

D15 The pupil's scale drawing

D16 (a) 14.2 cm   (b) 2.9 cm

D17 (a) 9.95   (b) 1.5 cm

D18 1 cm represents 20 m.

D19 1 cm represents 40 m.

E Ratios (p 163)

E1 (a) 1:50   (b) 1:100
(c) 1:500   (d) 1:2000

E2 (a) 1:5000   (b) 1:1000

E3 (a) 1:10 000   (b) 1:50 000
   (c) 1:20 000   (d) 1:100 000

E4 1:5000

E5 (a) 2 m   (b) 40 m   (c) 2.5 m

E6 (a) 25 m   (b) About 160 m
   (c) About 110 m

E7 (a) 1:25   (b) 1:50
   (c) 1:20   (d) 1:20 000

What progress have you made? (p 164)

1 (a) B   (b) 2

2 $a = 3$ cm, $b = 10.5$ cm

3 (a) 12 km   (b) 12 cm

4 (a) 1:1000   (b) 1:5000
   (c) 1:250   (d) 1:50 000

Practice booklet

Sections A, B and C (p 72)

1 The pupil's enlargement

2 (a) The angles of corresponding vertices are the same.

   (b) It is an enlargement.

3 (a) The angles of corresponding vertices are the same.
   (b) It is not an enlargement.

4 (a) $\frac{1}{2}$ (b) 2

5 (a) $\frac{1}{3}$ (b) 3

6 A enlargement of B   scale factor $\frac{3}{2}$ or $1\frac{1}{2}$
   B enlargement of A   scale factor $\frac{2}{3}$
   C enlargement of D   scale factor $\frac{3}{2}$ or $1\frac{1}{2}$
   C enlargement of E   scale factor $\frac{3}{4}$
   D enlargement of C   scale factor $\frac{2}{3}$
   D enlargement of E   scale factor $\frac{1}{2}$
   E enlargement of C   scale factor $\frac{4}{3}$ or $1\frac{1}{3}$
   E enlargement of D   scale factor 2

7 $a = 2$ cm, $b = 8$ cm, $c = 3$ cm, $d = 6$ cm, $e = 1$ cm, $f = 3$ cm

8 A enlargement of C   scale factor 2
   B enlargement of E   scale factor $\frac{1}{4}$
   B enlargement of G   scale factor $\frac{1}{3}$
   B enlargement of H   scale factor $\frac{1}{2}$
   C enlargement of A   scale factor $\frac{1}{2}$
   E enlargement of B   scale factor 4
   E enlargement of G   scale factor $\frac{4}{3}$ or $1\frac{1}{3}$
   E enlargement of H   scale factor 2
   G enlargement of B   scale factor 3
   G enlargement of E   scale factor $\frac{3}{4}$
   G enlargement of H   scale factor $\frac{3}{2}$ or $1\frac{1}{2}$
   H enlargement of B   scale factor 2
   H enlargement of E   scale factor $\frac{1}{2}$
   H enlargement of G   scale factor $\frac{2}{3}$

Pupils may not find all the fractional scalings. Shapes D and F are not related by scaling to any other shapes.

Section D (p 75)

1 (a) 7.9 cm   (b) 790 m

2 (a) 680 m   (b) 500 m
   (c) 1300 m   (d) 380 m

3 The pupil's description of method

4 12 cm

5 (a) 114 m   (b) 54 m   (c) 78 m

6 (a) 8 cm   (b) 100 cm   (c) 0.8 cm

7 Scale factor 5

Section E (p 76)

1 (a) 1:25 000   (b) 1:500 000
   (c) 1:100

2 (a) 1:50 000   (b) 1:2500
   (c) 1:2000

3 (a) 1:10 000   (b) 1:2000

23 Approximation and estimation

Essential

Sheet 217

Practice booklet pages 77 and 78

𝔸 **Rough estimates** (p 165)

Sheet 217

Seats

◊ Ask pupils to think about how they could get a rough idea of the number of seats in each hall on the resource sheet. Don't ask anyone to give their answer at this stage; ask them to discuss their method with a partner. Then get each pair to find an estimate.

Now ask pupils to describe how they got their estimate and compare the methods and the results. For example, some may split the seats into blocks of 100. Lead pupils (if they haven't already got there themselves) to the idea of using approximations for the number of rows and the number of seats in each row:

Century Hall: roughly 20 rows with roughly 30 seats per row (estimated 600, actual number 22 × 28 = 616)

Millennium Hall: roughly 20 rows with roughly 40 seats per row (estimated 800, actual number 19 × 39 = 741)

You can also ask how in the second case they can know in advance that the estimate will be lower than the actual number.

◊ When doing, for example, 60 × 300, pupils usually see that it's 18 followed by three noughts. 'Adding noughts' works with whole numbers only, so help pupils to see that it comes about as a result of multiplying by 10 several times.

B Rounding to one significant figure: whole numbers (p 166)

◊ The most significant figure is the most important figure in telling us the size of the number. Obviously with whole numbers the most significant figure is the first figure.

The process of rounding to one significant figure can be thought of as in this example.

3704   Identify the most significant figure: 3 (thousands).
             The number is between 3000 and 4000.
             It is nearer to **4000**.

A slight complication arises when the most significant figure is a 9 and rounding up is necessary.

9531   Identify the most significant figure: 9 (thousands).
             The number is between 9000 and 10 000.
             It is nearer to **10 000**.

C Estimation: whole numbers (p 167)

D Rounding to one significant figure: decimals (p 167)

◊ It may help to show that a whole number like 3704 can be thought of as …0003704. Reading from left to right, the most significant figure is the first non-zero figure you come to.

The same idea then applies to decimals.

0.03704   Identify the most significant figure: 3 (second d.p.).
                The number is between 0.03 and 0.04.
                It is nearer to **0.04**.

◊ Complications arise when the first significant figure is a 9 and rounding up is necessary. (These first occur in question D3.)

0.953   Identify the most significant figure: 9 (first d.p.).
              The number is between 0.9 and 1.
              It is nearer to **1**.

0.0962   Identify the most significant figure: 9 (second d.p.).
               The number is between 0.09 and 0.1.
               It is nearer to **0.1**.

In both cases the second line of the argument is likely to cause difficulty. Using a number line may help.

0.5	0.6	0.7	0.8	0.9	What's next?
0.05	0.06	0.07	0.08	0.09	What's next?

E Multiplying decimals (p 168)

◊ Emphasise that every time one of the numbers is multiplied/divided by 10, so is the result.

◊ It is not necessary to start every time from a tables fact. For example, working out 0.03 × 400 can start from 3 × 400.

◊ A common error is '0.3 × 0.2 = 0.6' (or similar). Even if pupils can see that 0.3 × 2 = 0.6, they still fall into the trap. An alternative way of looking at it is through area.

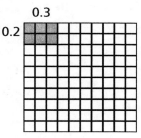

F Estimation (p 169)

A Rough estimates (p 165)

A1 (a) 60 (b) 600 (c) 200
(d) 2000 (e) 20 000 (f) 120
(g) 1200 (h) 12 000 (i) 1200
(j) 1200 (k) 1000 (l) 9000
(m) 1500 (n) 1600 (o) 16 000

A2 (a) 600 (b) 2000 (c) 1200
(d) 2400 (e) 2400

A3 2100

A4 800

A5 (a) 8000 (b) 6000 (c) 21 000
(d) 20 000 (e) 40 000

B Rounding to one significant figure: whole numbers (p 166)

B1 (a) 500 (b) 7000 (c) 500
(d) 1000 (e) 80 (f) 4000
(g) 700 (h) 3000 (i) 9000
(j) 600

B2 The pupil's headlines, including
(a) 300 (b) 200 (c) £900
(d) 7000

B3 (a) 50 000 (b) 80 000
(c) 60 000 (d) 50 000
(e) 10 000

B4 The pupil's headlines, including
(a) 20 000 (b) £60 000 (c) 700 000

B5 (a) 100 (b) 1000 (c) 7000
(d) 40 000 (e) 90 000

C Estimation: whole numbers (p 167)

C1 (a) 2000 (b) 12 000 (c) 15 000
(d) 7200 (e) 5000 (f) 6000
(g) 36 000 (h) 72 000

C2 14 000 m²

C3 (a) $8000 \, \text{m}^2$ (b) $8000 \, \text{m}^2$
    (c) $12\,000 \, \text{m}^2$ (d) $4000 \, \text{m}^2$
    (e) $30\,000 \, \text{m}^2$ (f) $64\,000 \, \text{m}^2$
    (g) $180\,000 \, \text{m}^2$ (h) $40\,000 \, \text{m}^2$

C4 £200

D Rounding to one significant figure: decimals (p 167)

D1 (a) 50   (b) 5   (c) 7
    (d) 20   (e) 80   (f) 0.4
    (g) 0.7   (h) 0.05   (i) 0.08
    (j) 0.02

D2 (a) 5000   (b) 400   (c) 6
    (d) 0.03   (e) 800   (f) 600
    (g) 0.007   (h) 1   (i) 300
    (j) 0.09

D3 (a) 1   (b) 0.09   (c) 0.1
    (d) 30   (e) 0.07   (f) $30\,000$
    (g) 0.3   (h) 2   (i) 0.009
    (j) 0.01

E Multiplying decimals (p 168)

E1 (a) 4.2   (b) 1.22   (c) 0.58
    (d) 30.2   (e) 30.8   (f) 2.08
    (g) 0.026   (h) 0.0007

E2 (a) $3 \times 4 = 12$
      $0.3 \times 4 = \mathbf{1.2}$
      $0.3 \times 0.4 = \mathbf{0.12}$
    (b) $4 \times 5 = 20$
      $4 \times 0.5 = \mathbf{2}$
      $0.4 \times 0.5 = \mathbf{0.2}$
    (c) $3 \times 5 = 15$
      $30 \times 5 = \mathbf{150}$
      $30 \times 0.5 = \mathbf{15}$
    (d) $2 \times 7 = 14$
      $0.2 \times 7 = \mathbf{1.4}$
      $0.02 \times 7 = \mathbf{0.14}$

    (e) $5 \times 0.5 = 2.5$
      $50 \times 0.5 = \mathbf{25}$
      $500 \times 0.5 = \mathbf{250}$
    (f) $40 \times 2 = \mathbf{80}$
      $400 \times 2 = \mathbf{800}$
      $400 \times 0.2 = \mathbf{80}$
    (g) $0.6 \times 3 = \mathbf{1.8}$
      $0.06 \times 3 = \mathbf{0.18}$
      $0.06 \times 30 = \mathbf{1.8}$
    (h) $0.5 \times 9 = \mathbf{4.5}$
      $0.5 \times 0.9 = \mathbf{0.45}$
      $0.5 \times 0.09 = \mathbf{0.045}$

E3 (a) 0.09 (b) 12   (c) 80   (d) 180
    (e) 40   (f) 0.04 (g) 300 (h) 64

E4 (a) 0.016   (b) 1.2   (c) 7
    (d) 1.2   (e) 140   (f) 1.6
    (g) 200   (h) 1.6

F Estimation (p 169)

F1 (a) 1.5   (b) 80   (c) 0.15 (d) 3.6
    (e) 50   (f) 2.4   (g) 0.36 (h) 72

F2 (a) $100 \, \text{m}^2$ (b) $0.8 \, \text{m}^2$ (c) $120 \, \text{m}^2$
    (d) $0.24 \, \text{m}^2$ (e) $0.06 \, \text{m}^2$ (f) $1.8 \, \text{m}^2$

F3 Rough estimate = 1600
    Gareth's answer is much too large.

F4 60

F5 $6 \, \text{m}^2$

F6 (a) $15 \, \text{m}^2$ (b) $12 \, \text{m}^2$

What progress have you made? (p 169)

1 (a) 400   (b) 8000   (c) 50 000

2 (a) 1600   (b) 18 000   (c) 320 000

3 (a) 40   (b) 0.5   (c) 0.08

4 (a) 24   (b) 0.06   (c) 2.8

Practice booklet

Sections A, B and C (p 77)

1 (a) 80   (b) 1800   (c) 240
   (d) 2500   (e) 21 000   (f) 3200
   (g) 30 000   (h) 2800

2 (a) 1200   (b) 12 000   (c) 14 000
   (d) 72 000

3

×	13	36	49	88
18	234	648	882	1584
27	351	972	1323	2376
64	832	2304	3136	5632
73	949	2628	3577	6424

4 2700

5 1500

6 (a) 400   (b) 8000   (c) 50 000
   (d) 100

7 (a) 300 m   (b) 8000 km (c) 700 g
   (d) 1000 litres

8 (a) 12 000 m^2 (b) £300   (c) £16 000

Sections D, E and F (p 78)

1 (a) 3.4   (b) 2.55   (c) 0.06
   (d) 0.3   (e) 240   (f) 0.49
   (g) 0.034   (h) 1.6   (i) 4.2
   (j) 240   (k) 0.025   (l) 27

2 (a) 40   (b) 500   (c) 0.04
   (d) 6   (e) 7   (f) 40
   (g) 8   (h) 0.04   (i) 0.07
   (j) 0.2   (k) 0.006   (l) 1

3 (a) 80 g   (b) 0.07 kg   (c) 5 g
   (d) 0.8 litre   (e) 0.3 litre   (f) 0.08 g
   (g) 0.006 g   (h) 500 kg   (i) 0.6 m
   (j) 5 km   (k) 6000 km   (l) 50 m

4 80 m^2

5 (a) 150 m^2 (b) 40 m^2 (c) 30 m^2
   (d) 60 m^2

6 Pupils may reasonably judge that an estimate is high or low, even when no (ii) response is given here.

   (a) (i) 600   (ii) low
       (iii) 736

   (b) (i) 400 000 (ii) high
       (iii) 372 710

   (c) (i) 3.6   (ii) low
       (iii) 3.6491

   (d) (i) 3   (ii) low
       (iii) 3.84

   (e) (i) 560   (ii) –
       (iii) 547.56

   (f) (i) 36   (ii) low
       (iii) 39.68

   (g) (i) 5400   (ii) –
       (iii) 5016

   (h) (i) 600   (ii) –
       (iii) 770

   (i) (i) 0.15   (ii) –
       (iii) 0.1368

   (j) (i) 0.032   (ii) high
       (iii) 0.029 64

   (k) (i) 7200   (ii) high
       (iii) 6907.8

   (l) (i) 1   (ii) –
       (iii) 1.058

㉔ Bearings

> **Essential**
> Angle measurer/protractor
> Sheets 202 to 205
>
> **Practice booklet** page 79

Ⓐ **Direction** (p 170)

> Angle measurer/protractor, sheet 202 (each pupil may need two copies)

◊ Before using the examples on the page, you could review compass directions: N, E, NE, etc. Relate each of them to an angle measured clockwise from N as zero, and emphasise that the angle corresponding to, say, NE is written as 045°.

The bearings shown on the page are 125° for the direction of the tanker and 250° for the direction of the ferry.

Note: Rathlin Island (sheet 202) is in Northern Ireland.

Ⓑ **On the moors** (p 172)

> Angle measurer/protractor, sheet 203

The area used for this section is part of Exmoor National Park.

B10 The bearing of a reverse direction is called a 'back bearing'. It is either 180° greater or 180° less than the bearing of the forward direction.

Ⓒ **Bearings jigsaw puzzle** (p 173)

> Angle measurer/protractor, sheets 204 and 205

◊ This puzzle gives plenty of practice in bearings and some scope for logical thinking.

𝔸 Direction (p 170)

A1 (a) 008°   (b) 032°

A2 to A6

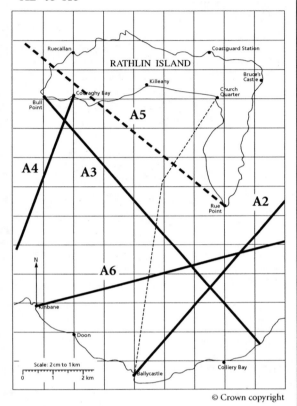

© Crown copyright

A5 On her left

A7 to A12

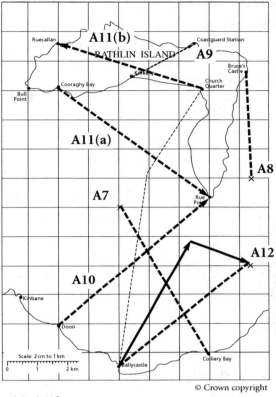

© Crown copyright

A9 062°

A10 049°

A11 (a) 127°   (b) 288°

A12 5.7 km on a bearing of 050°

A13 and A14

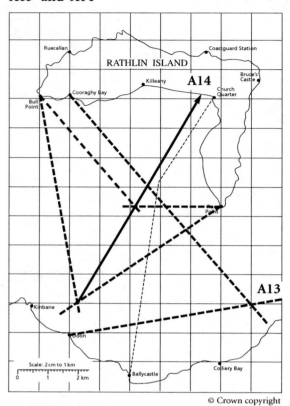

© Crown copyright

🅱 On the moors (p 172)

B1 12 cm

B2 (a) 18 cm (b) 11.2 cm (c) 6.4 cm
(d) 19.6 cm (e) 3.2 cm

B3 4 km

B4 (a) 2 km (b) 3.75 km (c) 1.8 km
(d) 2.6 km (e) 1.35 km

B5 to B9

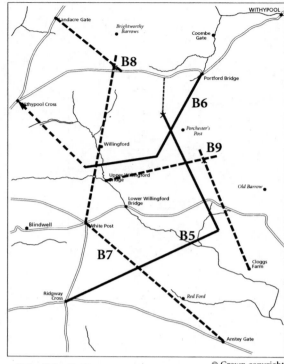

© Crown copyright

B5 (b) 0.6 km

B6 (b) 315°

B7 (a) 312° (b) 3 km

B10 (a)

RC to CF	079°	CF to RC	259°
CG to OB	161°	OB to CG	341°
PP to W	258°	W to PP	078°
WC to WP	151°	WP to WC	331°
RF to OB	036°	OB to RF	216°

(b) The difference between them is 180°.

ℂ Bearings jigsaw puzzle (p 173)

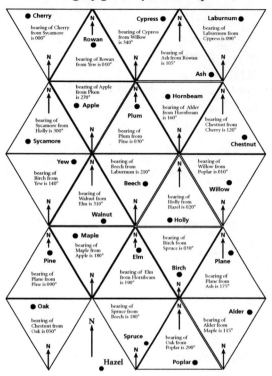

What progress have you made? (p 173)

1 (a) 124°   (b) 281°

2 and 3

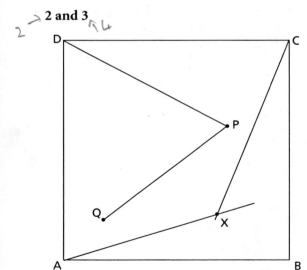

Practice booklet

Section A (p 79)

1 (a) 090°   (b) 058°   (c) 000°

(d)

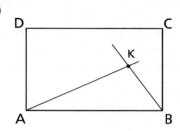

(e) 110°   (f) 25.6 km

2 (a) 310°   (b) 190°

3 (a) 015°   (b) 330°   (c) 105°

(d) 195°   (e) 240°

Using equations

This unit introduces equations involving brackets and equations of the type $ax - b = cx \pm d$. Where subtraction is involved, pupils can no longer think of the equations as representing a real-life balance. They need to think of the two sides of the equation as representing the same number. The emphasis is on doing the same thing to both sides of the equation.

T	p 174 **A** Walls	Revision of previous work
T	p 176 **B** Both sides	Equations with the unknown on both sides, some involving brackets
T	p 177 **C** Think of a number	Translating number puzzles into equations, including those with brackets
T	p 178 **D** Subtracting	Solving equations with subtracted numbers such as $2x + 4 = 3x - 5$ and $2x - 4 = 3x - 5$
T	p 180 **E** Equations from pictures	Forming equations from situations described in pictures or words

Optional

The computer program SOLVE (part of the Slimwam 1 pack available from ATM, tel 01332 346599)

Practice booklet pages 80 to 83

𝔸 **Walls** (p 174)

In this section the equations involve addition only. So each equation can be thought of concretely as equivalent to a balance with real weights and objects.

◊ Pupils in groups can try to complete the two incomplete addition walls. They should find that the second wall is not so easy as the first. By the end of your discussion, pupils should be able to apply the method used in A1.

B **Both sides** (p 176)

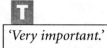

◊ The emphasis in the introduction needs to be on solving an equation by 'doing the same thing to both sides' rather than 'subtracting the same thing'. 'Doing the same thing' will be built on in later sections.

◊ Include the idea of checking in your introduction. One method of setting out a check might be:

When $x = 11$, $21 + x = 21 + 11 = 32$ and
$2x + 10 = 2 \times 11 + 10 = 32$, so the solution checks.

Where the equation arises from a problem, encourage pupils to also check that the original problem is solved. In this case the equation arises from a 'wall' problem. The value $x = 11$ gives the wall

The two bricks in the second top row are indeed the same.

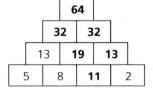

C **Think of a number** (p 177)

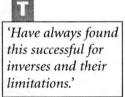

◊ Groups of pupils could discuss solving Jim's and Ollie's puzzles, and then report back to the whole class.

They need to realise why they cannot use the inverse approach when solving Ollie's puzzle. Again, the discussion needs to bring out the need to 'do the same thing to both sides of the equation'.

◊ Encourage pupils to check their solution fits their equation *and* the original number puzzle.

D **Subtracting** (p 178)

> Optional:
> The computer program SOLVE (part of the Slimwam 1 pack available from ATM, tel 01332 346599) has been found very useful for this type of work.

◊ The idea of a balance can only be applied to equations such as $5a - 9 = 4a + 1$ in the most artificial way, since we cannot easily show the '$- 9$' on the left-hand side. However, the balance idea has emphasised 'doing the same thing to both sides' and this approach is used here.

◊ Pupils often find equations where there is a subtraction on both sides more difficult (this type is in the second box). Questions D1 to D7 involve subtraction on one side of the equation only so this section can be split into two teaching sessions.

◊ The introductory equations could be discussed in groups. Then each could report back to the class in a class discussion.

◊ If we wish to isolate the $5a$ on the left-hand side of $5a - 9 = 4a + 1$, then we must add 9 to both sides.
Pupils' experiences in 'Think of a number' (*Book S1* unit 13) should help them see that, in order to get back from $5a - 9$ to a, they have first to undo the $- 9$.

In solving equations of this type, pupils often think that they can 'take off 9' from $5a - 9$, leaving just $5a$. It may help to emphasise 'How do you undo "subtract 9"? You add 9.'

'Important point to make.'

If the point does not naturally arise in discussion, bring out that there is no one correct thing to do first. You could add 9 to both sides, or subtract 1 from both sides with equal validity. However, not all first steps are equally useful!

You may wish to suggest to pupils that they tackle the x's first, so that later, when solving equations where x's are subtracted, they do not get left with any $- x$'s. Other teachers prefer pupils to first get rid of all subtraction signs.

◊ There are of course innumerable pitfalls when solving equations. Another common mistake is to, say, take d off $d - 2$, and leave just 2.

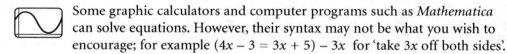

Some graphic calculators and computer programs such as *Mathematica* can solve equations. However, their syntax may not be what you wish to encourage; for example $(4x - 3 = 3x + 5) - 3x$ for 'take $3x$ off both sides'.

E Equations from pictures (p 180)

Many pupils will find these questions hard. There will be further opportunity to practise the skills involved in later work in *Book S3*.

Ⓐ Walls (p 174)

A1 (a) $7 + x$ (b) $x + 5$
 (c) $12 + 2x$ (d) $12 + 2x = 88$
 (e) $x = 38$ and completed wall

A2 (a) $x = 9$ (b) $x = 28$ (c) $x = 18$

A3 (a) $a = 11$ and completed wall
 (b) $p = 17$ and completed wall
 (c) $m = 3$ and completed wall
 (d) $t = 7$ and completed wall

A4 (a)

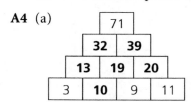

 (b)

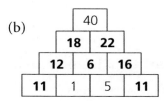

*A5 (a)

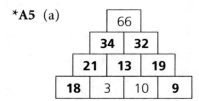

 (b)

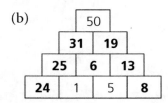

Ⓑ Both sides (p 176)

B1 (a) $x = 11$ (b) $x = 3$
 (c) $a = 7$ (d) $g = \frac{1}{2}$
 (e) $x = 10$ (f) $y = 1$
 (g) $p = 6$ (h) $e = 25$
 (i) $x = 3\frac{1}{2}$ (j) $t = 0$

B2 (a) The pupil's working leading to $m = 11$
 (b) When $m = 11$,
 $7(m + 3) = 7 \times (11 + 3) = 7 \times 14 = 98$
 $3m + 65 = 3 \times 11 + 65 = 33 + 65 = 98$

B3 (a) $2n + 6$
 (b) $n = 4$
 (c) When $n = 4$,
 $2(n + 3) = 2 \times (4 + 3) = 2 \times 7 = 14$
 $n + 10 = 4 + 10 = 14$

B4 (a) $x = 4$ (b) $w = 7$
 (c) $t = 1$ (d) $d = 3$
 (e) $g = 11$ (f) $k = 15$
 (g) $h = 2$ (h) $r = 3$

*B5 (a)

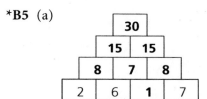

 (b)

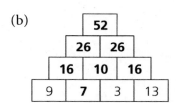

Ⓒ Think of a number (p 177)

C1 Call Harry's number n.
 $6n + 20 = 10n$
 $n = 5$

C2 (a) $9n + 18 = 11n$
 $n = 9$
 (b) $4n + 100 = 8n$
 $n = 25$

C3 $d = 4$

C4 (a) $5(n + 4) = 7n$
 $n = 10$
 (b) $6(n + 12) = 8n$
 $n = 36$

C5 $n = 4$ and the pupil's check

C6 $5(n + 2) = 3n + 28$
$$n = 9$$

C7 $3(n + 7) = 7n + 5$
$$n = 4$$

Ⓓ Subtracting (p 178)

D1 $t = 9$

D2 $z = 2$

D3 If you take 2 from $4d - 2$ you do not get rid of the 2, you get $4d - 4$.

One solution is (could be ordered differently)
$$4d - 2 = 2d + 6$$
$$4d = 2d + 8$$
$$2d = 8$$
$$d = 4$$

D4 (a) $b = 3$ (b) $y = 4$
(c) $h = 15$ (d) $k = 5$
(e) $m = 5$ (f) $d = 20$
(g) $f = 6$ (h) $q = 2$

D5 (a) $3x + 18$
(b) $x = 10$
(c) When $x = 10$,
$3(x + 6) = 3 \times (10 + 6) = 3 \times 16 = 48$
$5x - 2 = 5 \times 10 - 2 = 50 - 2 = 48$

D6 (a) $t = 4$ (b) $h = 8$
(c) $j = 7$ (d) $a = 18$
(e) $q = 12$ (f) $e = 5$

D7 (a) $4(n - 3) = 2n$
$$n = 6$$
(b) $7n - 100 = 3n$
$$n = 25$$

D8 $z = 3$

D9 If you take $3s$ from $3s - 2$ you do not get 2, you get -2.

One solution is (could be ordered differently)
$$3s - 2 = 5s - 8$$
$$3s + 6 = 5s$$
$$6 = 2s$$
$$s = 3$$

D10 (a) $n = 3$ (b) $y = 3$
(c) $t = 1$ (d) $m = 4$
(e) $x = 2$ (f) $c = 3$

D11 (a) $y = 5$ (b) $g = 5$
(c) $u = 10$ (d) $h = 8$

D12 (a) $b = 2$ (b) $y = 5$
(c) $h = 4$ (d) $k = 4$
(e) $m = 10$ (f) $d = 6$

D13 15

Ⓔ Equations from pictures (p 180)

E1 11 cubes in each column

E2 (a) $1400 (b) $5400 each

E3 Each coach holds 35 people.

E4 (a) $2n - 16$ (b) $4n - 50$
(c) $2n - 16 = 4n - 50$
$$n = 17$$
There are 17 mints in each packet.

E5 (a) $10(w - 2)$ (b) 10 woodscrews

E6 Jenny and Bob are both 17.
Great-uncle Fred is 87.

What progress have you made? (p 182)

1 (a) $x = 13$ (b) $x = 14$

2 (a) $x = 3$ (b) $y = 9$
(c) $h = 3$

3 (a) $g = 10$ (b) $f = 7$

  (c) $d = 7 \; 12$

4 (a) $6n - 52 = 2n$

                $n = 13$

  (b) $4(n - 6) = 2n - 10$

                 $n = 7$

Practice booklet

Section A (p 80)

1 (a) $x = 5$

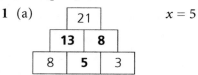

  (b) $x = 6$

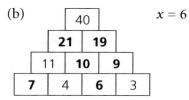

2 (a) $r = 10$

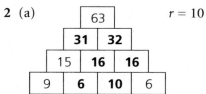

  (b) $p = 4$

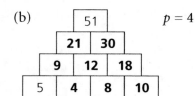

3 (a)   (b)

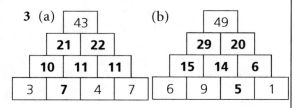

4 (a)   (b)

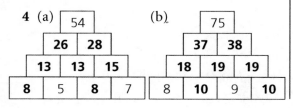

Section B (p 81)

1 (a) $a = 1$ (b) $x = 12$ (c) $y = 3\frac{1}{2}$

  (d) $p = 4$ (e) $t = 3$ (f) $x = 5\frac{1}{2}$

2 (a) $5n + 35$ (b) $n = 3$

3 (a) $x = 2$ (b) $y = 10$ (c) $z = 4$

  (d) $g = 2\frac{1}{2}$ (e) $h = 1$ (f) $p = 1$

Section C (p 81)

In these, any letter maybe used for the unknown number.

1 (a) $7x + 15 = 10x$ $x = 5$

  (b) $4(x + 8) = 6x$ $x = 16$

2 $4x + 9 = 3(x + 5)$ $x = 6$

3 $4(x + 7) = 8(x + 2)$ $x = 3$

Section D (p 82)

1 $z = 4$ with the pupil's check

2 (a) $b = 5$ (b) $t = 9$ (c) $d = 10$

3 (a) $y = 1$ (b) $x = 2$ (c) $h = 5$

4 (a) $x = 9$ (b) $x = 9$ (c) $t = 6$

  (d) $y = 11$ (e) $x = 7$ (f) $y = 3$

5 $6(x - 13) = 2x + 14$ $x = 23$

Section E (p 82)

1 15 sweets

2 6 pupils

3 4 apples

4 (a) $15(f - 3)$

  (b) 15 pieces

5 Alicia and Matthew are 13 years old. Their teacher is 32 years old.

26 Distributions

This unit revises earlier work on median, mode, mean and range, and introduces stem-and-leaf tables and the distinction between discrete and continuous data.

Practice booklet pages 84 to 88

Ⓐ Median, range and mode (p 183)

◊ These questions are for revision.

Ⓑ Stem-and-leaf tables (p 184)

◊ Measuring pulse rates provides suitable data for a stem-and-leaf table, and the data is relatively easy to collect. When pupils are sure that they have found their pulse, you can start and finish a one-minute interval. The numbers of beats per minute are recorded on a class table similar to this.

```
4 | 7 ◄──────────────  This shows a pulse rate
5 | 6 3 8 2              of 47 beats per minute.
6 | 3 7 1 0 5
7 | 2 6 3 2
```

When all the data has been entered, the table is rewritten with the units digits in order of size.

```
4 | 7
5 | 2 3 6 8
6 | 0 1 3 5 7
7 | 2 2 3 6
```

Working on squared paper helps to keep the digits in line.

◊ Discuss how to describe the modal group (60–69 in the example above) and how to find the median and range from a stem-and-leaf table.

ⓒ **Mean** (p 185)

This section revises earlier work on the mean. Questions C4 and C5 informally introduce the use of an assumed mean.

Ⓓ **Discrete and continuous data** (p 187)

◊ There is no need to make too much of the distinction between discrete and continuous data. In practice, continuous measurements are rounded to a certain degree of accuracy, and in that form are actually discrete.

◊ Throughout this unit a boundary value is put into the upper interval. Where to include a boundary value can be indicated by using inequalities to describe intervals, for example $2 \leq w < 3$. You may wish to introduce this notation now. (It is dealt with in a later unit.)

Ⓔ **Summarising and comparing data** (p 190)

◊ The mode, median or mean may or may not be a useful measure of average: it depends on the data. Question E2 gives an example where none of the three gives a good 'average' because it is debatable whether there is such a thing as a 'typical' value for this set of data.

◊ Scatter graphs are briefly introduced. They are dealt with more fully in *Book S3*.

Ⓐ **Median, range and mode** (p 183)

A1 Median 137 cm, range 36 cm

A2 (a) 46 kg   (b) 20 kg

A3 Median 141 cm, range 34 cm

A4 (a) 44 cm   (b) 32 cm

    (c) The boys' range is larger. It shows that the boys' heights are more spread out than the girls'.

A5 3

A6 (a) 39   (b) 2

A7 3

A8 A5: 3   A6: 1   A7: 3

Ⓑ **Stem-and-leaf tables** (p 184)

B1 (a) 48   (b) 27   (c) 73   (d) 70–79

B2 Paper 1

2	8
3	8 9
4	1 3 4 5 9 9
5	0 4 4 4 5 8
6	0 1 2 3 3 6 6 6
7	0 0 2 9
8	0 2 2

Paper 2

2	8 9
3	4 5 7 8 8 9
4	0 1 2 4 4 7 8 9
5	0 1 2 4 5 6 8
6	1 2 3 4 9
7	0
8	3

Paper 2 seems to have been harder. The median marks are 59 on paper 1 and 48.5 on paper 2. Most marks on paper 1 are in the 40–70 range, but on paper 2 they are in the 30–60 range.

B3 (a)

1.	7 9 9
2.	3 5 5 6 7 8 8
3.	0 1 2 3 3 4 7 8
4.	0 1

(b) 2.4 kg   (c) 2.9 kg

ℂ **Mean** (p 185)

C1 3.1 kg

C2 Species A: mean 6.8 cm, range 2.9 cm
Species B: mean 7.7 cm, range 5.0 cm
The worms in species B are longer on average. They are also more spread out in length.

C3 Engine A: mean 12.7, range 6.7
Engine B: mean 15.1, range 3.8
Engine B has higher fuel consumption than engine A. The fuel consumptions of B are less spread out.

C4 (a) 6 cm   (b) 166 cm

C5 35.75

𝔻 **Discrete and continuous data** (p 187)

D1 (a) Variety A

Number of peas	Frequency
0–4	2
5–9	7
10–14	10
15–19	4
20–24	2

Variety B

Number of peas	Frequency
0–4	1
5–9	14
10–14	9
15–19	1
20–24	0

(b)

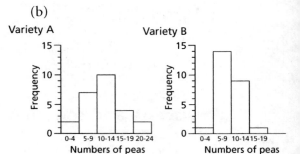

(c) 10–14   (d) 5–9   (e) A

D2 (a) About 3.7–3.8 kg

(b) Because you cannot see from the graph what the smallest and largest weights are, only the intervals they are in.

D3 (a) Girls

Handspan in cm	Frequency
16–17	2
17–18	3
18–19	7
19–20	4
20–21	3
21–22	1

Boys

Handspan in cm	Frequency
16–17	1
17–18	1
18–19	2
19–20	3
20–21	3
21–22	5
22–23	5

(b)

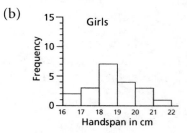

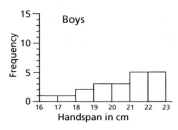

(c) The girls' handspans show many in the middle of the range with few very large or small. The boys' handspans are mostly large, with very few small.

D4 (a) He may be right. The graph shows that the longest journey is in the interval 20–25 minutes.

(b) We do not know the shortest and longest journey times.

(c) 5–10 minutes

(d) 40%   (e) 15%

D5 (a) 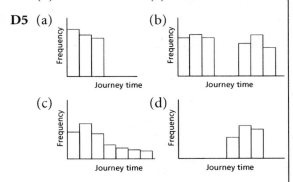 (b)

(c) (d)

E Summarising and comparing data (p 190)

E1 (a) £17k. It does not give a good idea of a typical salary.

(b) Median £11k, mode £9k. The median gives a better idea of a typical salary.

E2 (a) Mean £21.5k, median £14k, mode £12k

(b) None of them gives a good idea of a 'typical' salary. (There is no 'typical' salary in this company.)

E3 Company A: mean £16k, range £20k
Company B: mean £16k, range £12k
The mean salary is the same for both companies.
The salaries in company B are less spread out (or closer together).

E4 Company C: mean £20k, range £20k
Company D: mean £17.5k, range £20k
Company C's salaries are higher on average but both companies' salaries are equally spread out.

E5 (a) The pupil's graph

(b) The after practice scores

(c) The before practice scores

(d) One pupil got worse, and stands out from the rest of the scores on the scatter graph.

What progress have you made? (p 192)

1 (a) 25   (b) 43   (c) 43

2

Weight in kg	Frequency
10–15	4
15–20	4
20–25	6
25–30	1

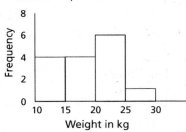

3 The apples from tree A are heavier on average. The apples from tree B are more spread out in weight.

4 The pupil's scatter graph

Practice booklet

Section A (p 84)

1 (a) 18 g   (b) 4 g

2 Median 51 mm, range 12 mm

3 (a) 5   (b) 6

4 (a) 25   (b) 5   (c) 7   (d) 6

5 Zoe's mice litter size has a smaller range.
Sam's mode is higher.
Both have the same median.

Section B (p 85)

1 (a) 51 minutes   (b) 27

   (c) 28 minutes   (d) 20–29 minutes

2 (a) (i) Soil A

```
2 | 1 4 5 6
3 | 0 2 2 4 6 8 9
4 | 0 1 1 3 5 7 7 8
5 | 1 6 7 9
6 | 0 2
```

            Soil B

```
2 | 6 8 9 9
3 | 0 3 3 5 5
4 | 1 2 2 4 6 6 7 8 8 8
5 | 0 0 1 1 3 4
```

     (ii) A: 41 cm   B: 28 cm

    (iii) A: 41 cm   B: 44 cm

    (iv) A: 40–49 cm   B: 40–49 cm

  (b) Plants grow more consistently in
soil B (smaller range in B).
The median is higher in B, although
some plants grow much taller in A.

3 (a)
```
1. | 5 8 9
2. | 0 2 4 6 6 7 8 9
3. | 1 1 2 4 4 7 9
4. | 0 1 2 2 3 4
```

  (b) 2.9 kg   (c) 3.1 kg

Section C (p 86)

1 3.25 kg

2 Mean £3.22, range £3.24

3 (a) Heat 1: mean 46.386 s, range 1.01 s
         Heat 2: mean 46.94 s, range 2.19 s

  (b) Heat 1 was closer (smaller range).
Heat 2 was slower (higher mean).

4 (a) Girls: mean 240,   range 217
         Boys: mean 215.6,   range 252

  (b) On average the boys wrote fewer
words per page.
There is less variation in the number
of words written by the girls.

  (c) 226.88

5 (a) 61.6%   (b) 61.2%

Section D (p 87)

1 (a)

No. of pens	Sasha's class	Hadeel's class
1–5	9	4
6–10	7	7
11–15	5	6
16–20	2	6

  (b)

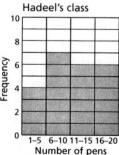

  (c) 1–5   (d) 6–10

2 (a)

Height (m)	Boys	Girls
1.20–1.30	1	3
1.30–1.40	5	6
1.40–1.50	5	4
1.50–1.60	4	2

(b) The pupil's frequency charts

(c) Overall the boys seem to be taller than the girls.

3 (a) 30–40   (b) 55

(c) The age is in the range 70–80.

Section E (p 88)

1 (a) 9 years   (b) 10.5 years

(c) $315 \div 24 = 13.125$ years

(d) The mode, or median, because most of the people on the coach are about this age.

2 (a) Mean £1.17 to the nearest penny
Median 20p
Mode 10p

(b) The pupil's choice with reason

3 (a) Ranges   British: 20 cm
Japanese: 19 cm

(b) Medians British: 147 cm
Japanese: 141.5 cm

Means   British: 147.3 cm
Japanese: 142.25 cm

(c) The Japanese students are smaller than the British students although the British students have a greater range in heights.

A sense of proportion

Essential

2 mm graph paper

Practice booklet pages 89 to 92

A Double up (p 193)

The objective here is to introduce the idea of direct proportion.

◊ Pupils can consider the problems in pairs or groups.

Some of the problems can clearly be solved by doubling (A, C, G) and some cannot (B and E). Problems D and F may need more discussion. After a bit of thought, most will accept that the height of an average four-year-old boy is not double the height of an average two-year-old boy (the average four-year-old boy is 101 cm tall). Problem F can be solved by doubling if one assumes that Joy will continue to cycle at the same rate but this may not be a reasonable assumption. We need to know things such as how fit Joy is and whether or not the 60 mile route is on the same sort of terrain as the 30 mile route. We can double to get a rough estimate but its accuracy depends on things we don't know about.

◊ Explain that when two quantities are in direct proportion, scaling one quantity up (or down) will result in the other being scaled up (or down) by the same factor and that the converse of this is true too.

A1 For problem B, an average 40-year-old man who is about 1.8 m tall weighs about 80 kg. A similar 20-year-old weighs about 73 kg.

Problems E and F may cause some discussion. It is a reasonable assumption that a fairly fit person's walking speed will not vary much over 5 miles so in problem E a reasonably good estimate can be found by halving. However, halving will produce a less accurate estimate in problem F as we know nothing about the routes. For example the route up Cracaval could be a long one with a shallow gradient or be much more difficult to negotiate than Sgurr Choinnich and in fact take much longer than 2 hours.

B Graphs (p 194)

| 2 mm graph paper |

◊ In the teacher-led discussion, two points need to be clearly made and understood. If quantities are in direct proportion
  - their graph will be a straight line sloping upwards
  - their graph will go through the point (0, 0)

Some will find the second point hard to grasp. You could ask pupils to try to draw a graph for two quantities in direct proportion that does not go through (0, 0).

C Conversion graphs (p 196)

| 2 mm graph paper |

◊ Once pupils are clear about the two correct rules that link k with m, they can use them to check their estimate for the first conversion (the rules give 12 miles equivalent to 19.2 km). Some pupils may be able to find a rule for m in terms of k.

D The unitary method (p 198)

This section reviews the unitary method, introduced in Unit 20 'Recipes'.

◊ Discuss why the unitary method can only be used for quantities that are in direct proportion.

◊ Remind pupils that they should not do any rounding until the end of a calculation.

Ⓐ Double up (p 193)

A1 A, D, E (and F if the mountains are very similar)

A2 A 15 grams   C $7\frac{1}{2}$ cm   D 8 tins

Ⓑ Graphs (p 194)

B1 (a) Graph B   (b) 4.4 cm or 4.5 cm

B2 (a) 400 g

(b) The pupil's graph through (0, 0), (3, 100), (6, 200), (9, 300), (12, 400)

(c) (i) About 130 g   (ii) About 330 g

Ⓒ Conversion graphs (p 196)

C1 (a) 37 km   (b) 28 miles
(c) 29 km   (d) 14 miles

C2 80 km with the pupil's method shown

C3 40 miles with the pupil's method shown

C4 (a) $22   (b) £27

C5 (a) $51   (b) £16   (c) $16

C6 The pupil's current conversion graph and questions

C7 (a) The pupil's graph through points (0, 0) and (50, 130)

(b) (i) 96 km²   (ii) 31 square miles

(c) $k = \frac{m}{50} \times 130$

Ⓓ The unitary method (p 198)

D1 (a) 55   (b) 385

D2 (a) 2.8 kg   (b) 33.6 kg

D3 £25

D4 136 euros

D5 17.9 kg

D6 2.1 kg

D7 19.2 kg

D8 126 m²

What progress have you made? (p 199)

1 20 minutes

2 (a) 5.3 pints   (b) 3.0 pints
(c) 3.4 litres   (d) 1.4 litres

3 36 kg

4 63.60 Swiss francs

Practice booklet

Sections A and B (p 89)

1 A, B, E

2 B   150 m²
C   16 km
E   1500 g or 1.5 kg

3 (a) Graph C   (b) 4.2 pints

Section C (p 91)

1 (a) €16 (b) €63 (c) €28 (d) €26

2 (a) The pupil's an answer between €96 and €98 with method

(b) The pupil's an answer between £46 and £48 with method

3 (a) The pupil's conversion graph going through the origin that matches 60 square yards with 50 square metres

(b) (i) 8 square metres
(ii) 48 square yards

(c) $m = \frac{y}{30} \times 25$

Section D (p 92)

1 45 g

2 50 kg

3 £0.60

4 16.5 kg

5 15.47 m

6 7210.5 yen

Fractions

Practice booklet pages 93 to 94

A Equivalent fractions (p 200)

◊ Establishing confidence with the strips can help avoid confusion when pupils carry out operations on fractions.

B Adding, subtracting and comparing (p 202)

◊ Mixed numbers and improper fractions were met in *Book S1*. You may need to revise them. The central idea here is selecting an appropriate common denominator.

C Multiplying (p 204)

◊ Two aspects of multiplying with fractions are considered here – so many lots of a fraction and a fraction of a quantity. The diagrams on page 205 show that these give the same result. Visualising (or sketching) diagrams is often a better way of dealing with fractions than applying 'rules' that are easily confused.

D Dividing a whole number by a fraction (p 206)

◊ Here, too, a diagrammatic approach is used.

Ⓐ Equivalent fractions (p 200)

A1 The pupil's fractions equivalent to $\frac{1}{2}$

A2 $\frac{3}{4} = \frac{9}{12}$

A3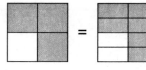

A4 (a) $\frac{1}{6} = \frac{2}{12}$ (b) $\frac{1}{4} = \frac{3}{12}$ (c) $\frac{5}{6} = \frac{10}{12}$

A5 (a) $\frac{1}{8} = \frac{2}{16}$ (b) $\frac{4}{5} = \frac{12}{15}$ (c) $\frac{2}{3} = \frac{12}{18}$

    (d) $\frac{3}{8} = \frac{12}{32}$ (e) $\frac{25}{40} = \frac{5}{8}$ (f) $\frac{16}{24} = \frac{2}{3}$

    (g) $\frac{18}{30} = \frac{3}{5}$ (h) $\frac{45}{75} = \frac{3}{5}$ (i) $\frac{1}{3} = \frac{4}{12}$

    (j) $\frac{1}{5} = \frac{6}{30}$ (k) $\frac{1}{7} = \frac{4}{28}$ (l) $\frac{1}{10} = \frac{5}{50}$

    (m) $\frac{3}{4} = \frac{15}{20}$ (n) $\frac{2}{9} = \frac{4}{18}$ (o) $\frac{5}{8} = \frac{20}{32}$

    (p) $\frac{4}{11} = \frac{24}{66}$

A6 $\frac{3}{8}, \frac{6}{16}$ and $\frac{9}{24}$ $\frac{2}{3}, \frac{8}{12}$ and $\frac{10}{15}$ $\frac{4}{5}, \frac{8}{10}$ and $\frac{16}{20}$

A7 (a) $\frac{2}{3}$ (b) $\frac{1}{2}$ (c) $\frac{3}{4}$ (d) $\frac{1}{2}$

    (e) $\frac{6}{7}$ (f) $\frac{2}{5}$ (g) $\frac{2}{7}$ (h) $\frac{4}{5}$

    (i) $\frac{2}{3}$ (j) $\frac{1}{5}$

A8 (a) $\frac{1}{3}$ (b) $\frac{4}{9}$ (c) $\frac{3}{5}$ (d) $\frac{2}{3}$

    (e) $\frac{1}{3}$ (f) Cannot be simplified

    (g) $\frac{4}{5}$ (h) $\frac{3}{4}$ (i) $\frac{3}{7}$ (j) $\frac{5}{12}$

Ⓑ Adding, subtracting and comparing (p 202)

B1 (a) $\frac{4}{5}$ (b) $\frac{3}{7}$ (c) $\frac{2}{3}$ (d) $1\frac{4}{9}$ (e) $1\frac{1}{4}$

B2 $\frac{3}{8}$

B3 (a) $\frac{5}{6}$ (b) $\frac{5}{8}$ (c) $\frac{7}{10}$ (d) $\frac{5}{16}$ (e) $\frac{2}{9}$

B4 (a) $\frac{11}{16}$ (b) $\frac{1}{2}$ (c) $1\frac{5}{12}$ (d) $1\frac{1}{2}$ (e) $1\frac{1}{6}$

B5 (a) $\frac{7}{12}$ (b) $\frac{1}{12}$

B6 (a) $\frac{11}{12}$ (b) $\frac{5}{12}$ (c) $\frac{5}{12}$ (d) $\frac{1}{12}$

B7 Because 12 is a multiple of 3 and of 4

B8 (a) The pupil's strip (Ten divisions are the most suitable.)

    (b) $\frac{7}{10}$ (c) $\frac{3}{10}$

    (d) The pupil's additions, subtractions and answers

B9 (a) The pupil's strip (Fifteen divisions are the most suitable)

    (b) $\frac{7}{15}$ (c) $\frac{13}{15}$

    (d) The pupil's additions, subtractions and answers

B10 (a) $\frac{9}{20}$ (b) $\frac{1}{6}$ (c) $\frac{19}{24}$ (d) $\frac{1}{6}$

    (e) $\frac{13}{30}$ (f) $\frac{1}{30}$ (g) $\frac{11}{30}$ (h) $\frac{1}{20}$ (or $\frac{2}{40}$)

    (i) $\frac{1}{24}$ (j) $\frac{17}{30}$

B11 (a) $\frac{23}{40}$ (b) $\frac{13}{20}$ (c) $\frac{3}{20}$ (d) $\frac{7}{30}$ (e) $\frac{5}{24}$

B12 (a) $\frac{3}{4}$ (b) $\frac{13}{24}$ (c) $1\frac{1}{12}$ (d) $\frac{13}{30}$ (e) $1\frac{1}{12}$

B13 (a) $\frac{3}{4}$ (or $\frac{15}{20}$)

    (b)

$\frac{2}{5}$	$\frac{3}{20}$	$\frac{1}{5}$
$\frac{1}{20}$	$\frac{1}{4}$	$\frac{9}{20}$
$\frac{3}{10}$	$\frac{7}{20}$	$\frac{1}{10}$

*B14 Towards the end of the fourth day. At the end of the fourth day it would have risen
$$\frac{10 - 3 + 10 - 3 + 10 - 3 + 10}{15} = \frac{31}{15} \text{ metres}$$

B15 (a) $\frac{3}{8}, \frac{8}{24} < \frac{9}{24}$ (b) $\frac{2}{3}, \frac{16}{24} > \frac{15}{24}$

    (c) $\frac{4}{5}, \frac{15}{20} < \frac{16}{20}$ (d) $\frac{5}{8}, \frac{24}{40} < \frac{25}{40}$

    (e) $\frac{4}{9}, \frac{27}{63} < \frac{28}{63}$ (f) $\frac{3}{8}, \frac{33}{88} > \frac{32}{88}$

    (g) $\frac{3}{5}, \frac{20}{35} < \frac{21}{35}$ (h) $\frac{9}{20}, \frac{80}{180} < \frac{81}{180}$

    (i) $\frac{7}{9}, \frac{27}{36} < \frac{28}{36}$ (j) $\frac{5}{7}, \frac{50}{70} > \frac{49}{70}$

B16 (a) $\frac{5}{9}$ (b) $\frac{3}{5}$ (c) $\frac{2}{5}$

Ⓒ Multiplying (p 204)

C1 (a) $\frac{6}{13}$ (b) $\frac{4}{5}$ (c) $\frac{9}{11}$ (d) $\frac{15}{16}$ (e) 1

C2 (a) $2\frac{2}{5}$ (b) $6\frac{2}{5}$ (c) $3\frac{3}{4}$ (d) $12\frac{1}{2}$ (e) $7\frac{1}{2}$

C3 (a) 8 (b) 8; they are the same

C4 (a)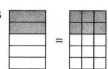

(b) $\frac{5}{3}$ (c) $1\frac{2}{3}$

C5 (a)

(b) $\frac{5}{4}$ (c) $1\frac{1}{4}$

C6 (a) $3\frac{2}{3}$ (b) $2\frac{1}{4}$ (c) $2\frac{2}{5}$ (d) $1\frac{7}{8}$

C7 (a)

(b) $\frac{4}{3}$ (c) $1\frac{1}{3}$

C8 (a)

(b) $\frac{6}{5}$ (c) $1\frac{1}{5}$

C9 (a) $2\frac{2}{3}$ (b) $5\frac{1}{4}$ (c) $1\frac{3}{5}$ (d) $4\frac{1}{8}$

Ⅾ **Dividing a whole number by a fraction** (p 206)

D1 6

D2 18

D3 20

D4 10

D5 12

D6 6

D7 (a) 24 (b) 4 (c) 8 (d) 3 (e) 18

What progress have you made? (p 206)

1 (a) $\frac{4}{5}$ (b) $\frac{3}{4}$

2 $\frac{5}{8}$, $\frac{35}{56} > \frac{32}{56}$

3 (a) $\frac{3}{8}$ (b) $\frac{13}{30}$ (c) $\frac{3}{20}$

4 (a) $1\frac{3}{5}$ (b) $1\frac{19}{36}$ (c) $1\frac{1}{12}$

5 (a) $11\frac{1}{4}$ (b) $5\frac{1}{4}$ (c) $4\frac{2}{3}$

6 (a) 72 (b) 12 (c) 10

Practice booklet

Section A (p 93)

1 $\frac{3}{12}, \frac{5}{20}, \frac{7}{28}$

2 $\frac{3}{5} = \frac{6}{10}$

3

4 $\frac{9}{15}, \frac{15}{25}$

5 (a) $\frac{1}{2}$ (b) $\frac{2}{7}$ (c) $\frac{5}{6}$ (d) $\frac{5}{14}$

6 The pupil's fractions equivalent to $\frac{1}{8}$

7 (a) $\frac{1}{5} = \frac{4}{20}$ (b) $\frac{5}{8} = \frac{15}{24}$
   (c) $\frac{3}{8} = \frac{12}{32}$ (d) $\frac{4}{9} = \frac{12}{27}$

Section B (p 93)

1 (a) $\frac{9}{12}$ (or $\frac{3}{4}$) (b) $\frac{5}{8}$ (c) $\frac{7}{10}$ (d) $\frac{13}{15}$

2 (a) $\frac{7}{15}$ (b) $\frac{4}{12}$ (or $\frac{1}{3}$) (c) $\frac{1}{20}$ (d) $\frac{17}{24}$

3 (a) $\frac{1}{10}$ (b) $\frac{7}{20}$

4 The pupil's fraction
   $\frac{1}{7}, \frac{1}{6}, \frac{1}{5}, \frac{1}{4}, \frac{5}{24}, \frac{7}{24}$ are possibilities.

5 $\frac{1}{5}, \frac{1}{4}, \frac{1}{3}, \frac{5}{9}, \frac{3}{5}$

Section C (p 94)

1 (a) $1\frac{3}{5}$ (b) $1\frac{1}{2}$ (c) $7\frac{1}{2}$ (d) $10\frac{1}{2}$

2

3 (a) $1\frac{5}{8}$ (b) $7\frac{1}{2}$ (c) $2\frac{2}{3}$ (d) $3\frac{1}{4}$

4 (a) $16\frac{1}{2}$ (b) $8\frac{1}{3}$ (c) $7\frac{1}{2}$ (d) $5\frac{1}{3}$

5 $2\frac{3}{4}$ cakes

6 (a) 10 (b) 18 (c) 14

Section D (p 94)

1 (a) 10 (b) 25 (c) 12 (d) 9

㉙ Constructions

Ⓐ Drawing triangles (p 207)

Compasses, angle measurers

◊ The ways of drawing triangles accurately correspond to conditions for congruence. You can point out that information in the form ASA, SAS and SSS defines a triangle uniquely (disregarding mirror images), and that two triangles with the same measurements in ASA, SAS or SSS form are therefore congruent.

A3 Part (a) does not define a single triangle but gives either of two triangles (see answers); so if two triangles have the same measurements in the pattern SSA it does not follow that they are congruent. Part (b) requires angle V to be calculated before drawing can start (so AAS does imply congruence).

◊ It is worth doing the constructions accurately on tracing paper or acetate, in order to check pupils' work easily and quickly.

Ⓑ Constructions with ruler and compasses (p 208)

Compasses, angle measurers

Pupils here have a taste of a 'pure' kind of drawing in which angle measurers and the graduations on rulers are not used (though angles are measured as an accuracy check in B2).

Ⓐ Drawing triangles (p 207)

A1 The pupil's drawings, with these measurements recorded

(a) Angle P = 80°, angle R = 62°, PR = 6.6 cm

(b) Angle A = 61°, angle B = 30°, angle C = 89°

(c) Angle L = 68°, KL = 8.9 cm, LM = 4.8 cm

(d) Angle D = 29°, angle F = 47°, DF =10.5 cm

A2 The pupil's accurate drawings

A3 (a) Two different triangles can be drawn with this information: in one case DE = 3.5 cm and in the other DE = 9.6 cm.

(b) Yes, angle V has to be calculated first, using the fact that the angles of a triangle add up to 180°.

Ⓑ Constructions with ruler and compasses (p 208)

B1 The pupil's construction; the perpendicular bisectors meet on the side AC.

B2 The pupil's construction; the angles of the large triangle are the same as the angles of the original triangle.

What progress have you made? (p 210)

1 The pupil's drawing

2 The pupil's drawing

3 The pupil's drawing

Practice booklet

Section A (p 95)

1 The pupil's drawings, with these measurements recorded

(a) Angle A = 53°, angle B = 34°, AB = 10.3 cm

(b) Angle R = 76°, PR = 3.5 cm, QR = 7.2 cm

(c) Angle L = 30°, angle M = 98°, angle N = 52°

(d) Angle E = 58°, angle F = 80°, EF = 7.7 cm

Section B (p 95)

1 The pupil's construction of an angle of 45°

2 (a) 60°

(b) The pupil's construction of an angle of 30°

③⓪ Division

This unit is to be done without a calculator.

Practice booklet pages 96 and 97

Ⓐ Simplifying divisions (p 211)

◊ You could explain the process concretely along these lines:

$\frac{120}{24}$ could mean, say, 120 apples shared between 24 people.

If there were only $\frac{1}{4}$ the number of apples and $\frac{1}{4}$ the number of people, then each person would still get the same number of apples.

This work has an obvious link with equivalent fractions.

Ⓑ Division with a decimal result (p 212)

◊ You may have to revise division by 10 and 100.

Ⓒ Dividing by a decimal (p 213)

◊ The effects of multiplying top and bottom by 10 or 100 can be shown by aligning decimal points and moving the figures the same number of places, like this

$$\frac{2.4}{0.03} \qquad \text{becomes} \qquad \frac{240.}{003.}$$

Ⓓ Rough estimates (p 214)

◊ Rounding to one significant figure is not always particularly useful. For example, it is easier to estimate $\frac{31.9}{0.38}$ by rounding to the 'nearest tables fact' ($32 \div 0.4$). However, all the examples in this section can be done by rounding to one significant figure.

Ⓐ Simplifying divisions (p 211)

A1 (a) 4   (b) 6   (c) 4   (d) 3
     (e) 3   (f) 5   (g) 8   (h) 6
     (i) 4   (j) 6

A2 (a) 20   (b) 5   (c) 40   (d) 300
     (e) 50   (f) 7   (g) 80   (h) 40
     (i) 4   (j) 900

A3 (a) 50   (b) 5   (c) 60   (d) 15
     (e) 25   (f) 50   (g) 300   (h) 20
     (i) 25   (j) 50

Ⓑ Division with a decimal result (p 212)

B1 (a) 0.2   (b) 0.2   (c) 0.3   (d) 0.8
     (e) 0.4   (f) 0.8   (g) 0.07 (h) 0.06
     (i) 0.4   (j) 0.3

B2 (a) 0.06   (b) 0.05   (c) 0.02
     (d) 0.02   (e) 0.004

B3 (a) 0.375   (b) 0.875

B4 0.6666…

B5 (a) 0.1111…
     (b) $\frac{2}{9} = 0.2222\ldots$
        $\frac{3}{9} = 0.3333\ldots$
        $\frac{4}{9} = 0.4444\ldots$ and so on

B6 $\frac{2}{7} = 0.2857142857\ldots$

     $\frac{3}{7} = 0.4285714285\ldots$

     $\frac{4}{7} = 0.5714285714\ldots$ and so on.

The same 'cycle' of six digits appears in every case, starting at a different point in the cycle for each fraction.

Ⓒ Dividing by a decimal (p 213)

C1 (a) 20   (b) 60   (c) 30   (d) 300
     (e) 8   (f) 7   (g) 4   (h) 0.6
     (i) 0.2   (j) 300

C2 (a) 40   (b) 70   (c) 200 (d) 40
     (e) 200 (f) 2   (g) 80   (h) 5
     (i) 300 (j) 400

C3 (a) $120 \div 0.5 = 240$ Answer: 240 bottles
     (b) $0.2 \div 4 = 0.05$ Answer: 0.05 kg
     (c) $3.0 \div 0.25 = 12$ Answer: 12 lengths
     (d) $6 \div 0.03 = 200$ Answer: 200 times
     (e) $2400 \div 0.4$ Answer: £6000.00 per kg

Ⓓ Rough estimates (p 214)

D1 (a) 20   (b) 20   (c) 200 (d) 50
     (e) 300 (f) 4   (g) 200 (h) 0.4
     (i) 0.08 (j) 500

D2 (a) 8   (b) 20   (c) 20   (d) 0.3
     (e) 20   (f) 40   (g) 200 (h) 0.3
     (i) 40   (j) 50

D3 (a) 20   (b) 30   (c) 50

What progress have you made? (p 214)

1 (a) 60   (b) 0.06 (c) 60   (d) 0.7

2 (a) 25   (b) 0.2   (c) 20   (d) 20

Practice booklet

Section A (p 96)

1 (a) 3   (b) 4   (c) 8   (d) 3
   (e) 5   (f) 6   (g) 9   (h) 7

2 (a) 6   (b) 110   (c) 700   (d) 6
   (e) 700 (f) 9   (g) 8   (h) 7

3 (a) 30   (b) 40   (c) 40   (d) 60
   (e) 50   (f) 80   (g) 600   (h) 25

Section B (p 96)

1 (a) 0.7   (b) 0.3   (c) 0.08 (d) 0.2
   (e) 0.06 (f) 0.05 (g) 0.04 (h) 0.045

2 (a) 0.04 (b) 0.06 (c) 0.09 (d) 0.003
   (e) 0.007 (f) 0.02 (g) 0.05 (h) 0.08

3 $\frac{1}{6} = 0.1666\ldots$ (recurring)

$\frac{2}{6} = 0.3333\ldots$ (recurring)

$\frac{3}{6} = 0.5$

$\frac{4}{6} = 0.6666\ldots$ (recurring)

$\frac{5}{6} = 0.8333\ldots$ (recurring)

Section C (p 97)

1 (a) 20 (b) 70 (c) 80 (d) 400
(e) 9 (f) 0.4 (g) 0.2 (h) 200

2 (a) 300 (b) 50 (c) 600 (d) 5
(e) 5 (f) 2500 (g) 400 (h) 80

3 15

4 17

Section D (p 97)

Other rough estimates may reasonably be given by the pupil.

1 (a) 0.08 (b) 20 (c) 140 (d) 30
(e) 60 (f) 20 (g) 50 (h) 0.06

2 (a) 16 (b) 40 (c) 40 (d) 20
(e) 120 (f) 0.03 (g) 1000 (h) 200

3 20

4 40

31 Indices

Practice booklet pages 98 to 101

A Factor trees (p 215)

One way to structure your introduction is given below.

◊ Ask pupils to continue the tree as far as they can. Discuss why it might be a good idea to disallow 1s in a factor tree. Ask how we know that the numbers at the end of the 'branches' must be prime if 1s are not allowed.

Pupils can then try to produce as many different factor trees as they can for 48. Discuss why the numbers at the ends of the branches are always the same.

Show how the factor tree helps to write 48 as a product of prime numbers $2 \times 2 \times 2 \times 2 \times 3$. There is no need to introduce index notation here – it is covered in section B.

◊ Discuss how the product could help you to find factors of 48 with questions such as

• How do we know straight away that 3 is a factor of 48?

• How do we know that 7 is not a factor of 48?

• How do we know that 6 is a factor of 48?

B Index notation (p 216)

Pupils have already seen index notation for squares and cubes and it is now extended to higher powers.

◊ There are a couple of problems given on page 2 to stimulate discussion.

By the end of your introduction pupils should be able to

• evaluate simple powers without a calculator

• evaluate higher powers with a calculator (giving an opportunity to use the power key if pupils have a good understanding of the concept)

- know that, say, 3^4 is a **power of 3**
- be able to read, say, 3^4 as 'three to the power four'
- be familiar with the words **index** and **indices**

ℂ **Powers of 10** (p 218)

𝔻 **Combining powers** (p 220)

◊ Unless you feel pupils are ready for it, there is no need to use these examples to introduce pupils to the formal rules for multiplying and dividing powers.

◊ Pupils can tackle these in groups and present their results to the rest of the class with reasons.

◊ At least to begin with, it will help pupils to write the multiplications out in full and then to write the results as powers of 2.

Pupils who feel that, say, $2^4 + 2$ can be simplified as a power of 2 could evaluate 2^4 to give $16 + 2 = 18$ which is not a power of 2.

For the divisions, it can help to evaluate each term and then express the result as a power of 2. For example $2^5 \div 2^2 = 32 \div 4 = 8 = 2^3$. Alternatively, pupils can think of a problem such as $2^5 \div 2^2$ as finding the missing expression in $2^2 \times ? = 2^5$ (just as, say, $24 \div 6$ can be thought of as finding the missing number in $6 \times ? = 24$).

𝔸 **Factor trees** (p 215)

A1 (a) The pupil's completed tree with 2, 2, 2, 5 on the ends of the 'branches'

(b) All trees for £40 end with 2, 2, 2, 5.

A2 The pupil's factor tree with 2, 2, 2, 3 on the ends of the 'branches'

A3 (a) (b)

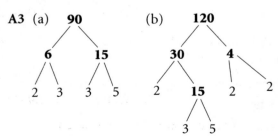

(c)

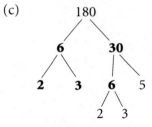

A4 (a) The pupil's factor tree with 2, 2, 3, 3 on the ends of the 'branches'

(b) $36 = \mathbf{2 \times 2 \times 3 \times 3}$

A5 (a) The pupil's factor tree with 2, 2, 7 on the ends of the 'branches'
28 = 2 × 2 × 7

(b) The pupil's factor tree with 2, 2, 3, 5, 5 on the ends of the 'branches' 300 = 2 × 2 × 3 × 5 × 5

(c) The pupil's factor tree with 3, 11 on the ends of the 'branches' 33 = 3 × 11

(d) The pupil's factor tree with 2, 2, 2, 2, 2, 2 on the ends of the 'branches'
64 = 2 × 2 × 2 × 2 × 2 × 2

A6 (a) 5　　(b) 2　　(c) 3　　(d) 2

A7 10, 20, 40, 50, 80

A8 (a) 70 = 2 × 5 × 7

(b) 2, 5, 7, 14

*A9 (a) 198 = 2 × 3 × 3 × 11

(b) 198 = (2 × 3) × (3 × 11)
= 6 × 33 so 6 is a factor

*A10 (a) 175 = 5 × 5 × 7

(b) The product does not contain 3 × 5.

ⓑ **Index notation** (p 216)

B1 C (5 × 5 × 5 × 5)

B2 B (4 × 4 × 4)

B3 (a) 5^3　　(b) 2^5　　(c) 7^4

B4 $2^4 = 16$　$3^2 = 9$　$4^3 = 64$　$2^3 = 8$

B5 (a) 32　(b) 81　(c) 16　(d) 64

B6 (a) 4　(b) 27　(c) 31　(d) 108

B7 (a) 243　　(b) 1296
(c) 1024　　(d) 3125

B8 (a) Greater　　(b) 6561

B9 2^7 is greater.

B10 (a) 216　　(b) 16 807

B11

[1]4	4	[2]1	■	[3]3
0	■	[4]9	7	2
■	■	6	■	■
[5]5	2	8		[6]6
0	■	[7]3	4	3

B12 (a) The pupil's factor tree with 2, 2, 2, 3, 3, 5 on the ends of the 'branches'

(b) $360 = 2^3 × 3^2 × 5$

B13 (a) $72 = 2^3 × 3^2$　　(b) $100 = 2^2 × 5^2$

(c) $392 = 2^3 × 7^2$

B14 (a) 2　　(b) 3　　(c) 3　　(d) 5

ⓒ **Powers of 10** (p 218)

C1

10^6	10 × 10 × 10 × 10 × 10 × 10	1 000 000	a million
10^5	10 × 10 × 10 × 10 × 10	100 000	a hundred thousand
10^4	10 × 10 × 10 × 10	10 000	ten thousand
10^3	10 × 10 × 10	1 000	a thousand
10^2	10 × 10	100	a hundred
10^1	10	10	ten

C2 A $(4 × 10^2) = 400$
B $(40 × 10^3) = 40 000$
C $(4 × 10^3) = 4000$

C3 (a) 50 000　　　(b) 23 000
(c) 200 000 000

C4 D

C5 　$2.5 × 10^3$
　= 2.5 × 1000
　= **2500**

C6 (a) 160　　(b) 1459　　(c) 345 000

C7 　$850 ÷ 10^2$
　= 850 ÷ 100
　= **8.5**

C8 (a) 5.9　　(b) 1.9174　　(c) 0.038

C9 2000 watts or 2 thousand watts

C10 3 000 000 tonnes or 3 million tonnes

C11 60 000 000 000 metres or 60 billion metres

C12 B (5 million megametres)

Ⅾ **Combining powers** (p 220)

D1 (a) 5^5　(b) 5^3　(c) 5^5　(d) 5^7

D2 (a) 3　　　(b) 3　　　(c) 3

D3 (a)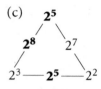

(b)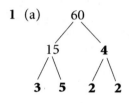

(c)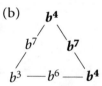

D4 (a) 3^2　(b) 3　(c) 3^3　(d) 3^2

D5 (a) 2^8　　　　(b) Not possible
　　(c) 2^3　　　　(d) Not possible

D6 (a) b^5　(b) n^7　(c) y^3　(d) x^4

D7 (a)

(b)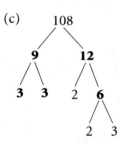

(c)

D8 (a) a　(b) n^2　(c) m　(d) y^2

What progress have you made? (p 221)

1 (a) The pupil's factor tree with 2, 2, 2, 2, 5 at the ends of its 'branches'

　(b) $80 = 2 \times 2 \times 2 \times 2 \times 5$

2 5^6

3 (a) 256　　(b) 2187　　(c) 625

4 (a) The pupil's factor tree with 2, 2, 2, 5, 5 at the ends of its 'branches'

　(b) $200 = 2^3 \times 5^2$

5 $500 = 2^2 \times 5^3$

6 (a) 30 000　　　　(b) 210
　(c) 5.6　　　　　　(d) 0.0425

7 (a) 2^7　　(b) 2^4　　(c) 2^2

8 (a) n^6　　(b) n^4　　(c) n

Practice booklet

Section A (p 98)

1 (a)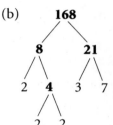

(b)

(c)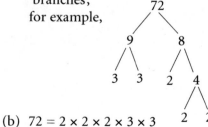

2 (a) The pupil's factor tree with 2, 2, 2, 3, 3 at the end of the 'branches', for example,

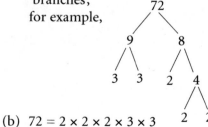

　(b) $72 = 2 \times 2 \times 2 \times 3 \times 3$

3 (a) $45 = \mathbf{3} \times 3 \times 5$
　(b) $75 = 3 \times \mathbf{5} \times 5$
　(c) $56 = 2 \times 2 \times \mathbf{2} \times 7$
　(d) $84 = 2 \times 2 \times \mathbf{3} \times 7$

4 (a) $120 = 2 \times 2 \times 2 \times 3 \times 5$

   (b) 3, 5, 15, 24 are factors of 120.

*5 (a) $156 = 2 \times 2 \times 3 \times 13$

   (b) $2 \times 2 \times 3 = 12$
     so $156 = 12 \times 13$
     so 12 is a factor.

Section B (p 98)

1 B

2 (a) 2^5 (b) 5^4 (c) 3^6

3 (a) 16   (b) 27   (c) 25   (d) 64

4 (a) The pupil's factor tree for 144, with 2, 2, 3, 2, 2, 3 on the ends of the 'branches'.

   (b) $144 = 2^4 \times 3^2$

5 (a) 256   (b) 729   (c) 16 384

   (d) 2401

6 (a) 17   (b) 432   (c) 174

   (d) 81

7 (a) 3^3 (b) 3×2^5 (c) $2^2 \times 7^2$

   (d) $2^2 \times 3^2 \times 5^2$

8 (a) $270 = 2 \times 5 \times 3^3$

   (b) $500 = 2^2 \times 5^3$

   (c) $396 = 2^2 \times 3^2 \times 11$

   (d) $675 = 3^3 \times 5^2$

Section C (p 99)

1 (a) 1000, 10^3

   (b) 10 000, 10^4

   (c) 1 000 000, 10^6

   (d) 100 000 000, 10^8

2 (a) 6000

   (b) 270 000

   (c) 40 000 000

   (d) 9 500 000

3 (a) 2400   (b) 17 500

   (c) 2620   (d) 862 000

4 (a) 0.62   (b) 1.51

   (c) 0.004 26   (d) 0.902

5 (a) 210 000 metres per hour

   (b) 11 800 000 watts

   (c) 10 200 000 000 metres

6 A and G, B and D, C and E
   F is the odd one out.

Section D (p 100)

1 (a) 4^5 (b) 4^5 (c) 4^7 (d) 4^6

2 (a) $3^2 \times 3^4 = 3^6$

   (b) $6^2 \times 6^7 = 6^9$

   (c) $8^5 \times 8 = 8^6$

3 (a) 5^2 (b) 5^2 (c) 5^5 (d) 5

4 (a) 3^2 (b) not possible

   (c) 3^7 (d) not possible

5 (a) b^6 (b) x^8 (c) y^5 (d) r^7

   (e) t^2 (f) m^2 (g) n^4 (h) z^3

6 (a) 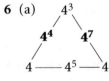   (b)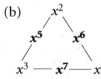

   (c)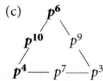

32 Rules and coordinates

Practice booklet   pages 101 to 102

Ⓐ Spot the function (p 222)

◊ Pupils can consider tables A to D in pairs or groups. In your discussion, revise the shorthand notation for the first three functions ($n \rightarrow n + 5$, $n \rightarrow 3n - 1$ and $n \rightarrow n^2 - 1$). Pupils may find it rather difficult to spot that the fourth function maps a number to its nearest 10. To help, you could give them some more pairs of values for the table.

◊ All the functions in A1 and A2 are linear except A2(c).

Ⓑ Graphing functions (p 223)

◊ Establish that the rule $y = 2x + 1$ gives rise to an infinite number of points that make a continuous straight line.

◊ Ask pupils if they can give examples of some functions that will give straight lines and some that will not give straight lines when graphed: there is no need to dwell on this at this stage. Question B2 is included as an example of a non-linear function.

Ⓒ Shapes within shapes (p 224)

C2 In part (d), possible comments are
  • triangle PQR is split into four congruent triangles
  • PQR is the same shape as each of the smaller triangles
  • the area of triangle PQR is four times the area of each smaller triangle

- the lengths of the edges of PQR are double the lengths of the edges of the smaller triangles
- each edge of the smaller triangle in the centre is parallel to one of the edges of PQR

C5 Pupils can try to explain their findings.

- It can help them see what is going on if they indicate the diagonals of the original quadrilateral by a dotted or coloured line.

  In the example below, we can argue using ideas of scaling that, since triangle ABD is an enlargement of triangle APS, PS is parallel to BD. By a similar argument QR is also parallel to BD and so QR must be parallel to PS. In the same way we can show that PQ is parallel to SR and so we will always have a parallelogram in the centre. This argument will also work for concave quadrilaterals.

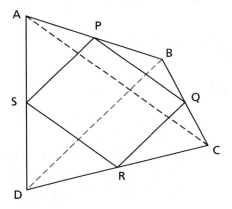

- A kite is impossible in the centre (except for the special cases of a square or rhombus) as it is not a parallelogram.
- The angles at the intersection of the diagonals of the original quadrilateral determine the interior angles of the final parallelogram (see diagram below).

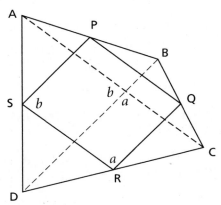

So, as long as the diagonals of the original quadrilateral cross at right angles, you will end up with a rectangle (special cases that give rectangles are kites and rhombuses). It can help pupils to start off with a rectangle and draw suitable quadrilaterals round it. However, very few pupils will spot this generalisation without some support and guidance from you.

- Using squared paper and calculating areas will suggest that the area of the parallelogram is half that of the original quadrilateral. A possible proof (for convex quadrilaterals) is shown below.

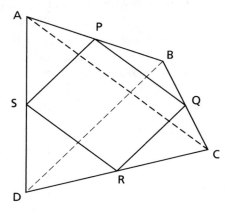

Area of APS = $\frac{1}{4}$ area of ABD

Area of BQP = $\frac{1}{4}$ area of BCA

Area of CRQ = $\frac{1}{4}$ area of CDB

Area of DSR = $\frac{1}{4}$ area of DAC

Adding the areas on the right-hand and left-hand sides gives

Area of ABCD – area of PQRS = $\frac{1}{4}$(area of ABD + area of CDB) +

$\frac{1}{4}$(area of BCA + area of DAC)

= $\frac{1}{4}$(area of ABCD) + $\frac{1}{4}$(area of ABCD)

= $\frac{1}{2}$(area of ABCD)

So the area of PQRS is $\frac{1}{2}$ the area of ABCD.

There is a similar, but slightly more 'fiddly', proof for concave quadrilaterals.

$\mathbb{D}$ Mid-points and coordinates (p 225)

$\mathbb{E}$ Coordinate code (p 226)

Ⓐ Spot the function (p 222)

A1 (a) 4 → **21**   (b) 3 → **1**
         1 → **6**   12 → **19**
        ‾2 → **‾9**   ‾1 → **‾7**
         2 → 11   7 → 9
        10 → 51   20 → 35

(c) 7 → **5**   (d) 18 → **12**
     0 → **12**   64 → **35**
    15 → **‾3**   ‾2 → **2**
    10 → 2   10 → **8**
    12 → 0   34 → 20

A2 (a) $n \rightarrow n - 4$ (b) $n \rightarrow 3n$
     (c) $n \rightarrow n^2$ (d) $n \rightarrow 2n + 3$

Ⓑ Graphing functions (p 223)

B1 (a) ‾4 → ‾8
         ‾2 → ‾2
          0 → **4**
          2 → **10**
          4 → **16**

(b) It will be a straight line.
(c) Yes, it is a straight line.

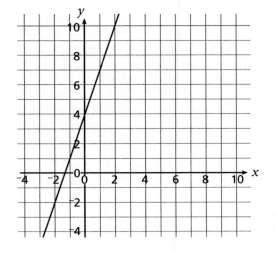

B2 (a)

x	y
‾4	16
‾2	4
0	0
2	4
4	16

(b) It will not be a straight line.
(c) No, it is not a straight line.

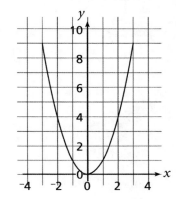

B3 (a)–(d)

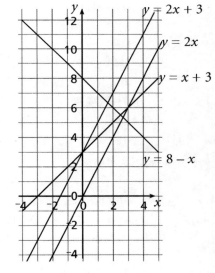

$y = 2x + 3$
$y = 2x$
$y = x + 3$
$y = 8 - x$

B4 The pupil's sketch of $y = 2$, a horizontal line through $(0, 2)$

C Shapes within shapes (p 224)

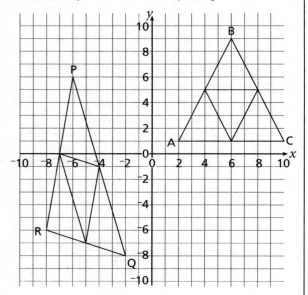

C1 (a) Points A, B and C plotted and joined
as above

   (b) Isosceles

   (c) Mid-point of AB marked; the
coordinates are (4, 5)

   (d) Mid-points of BC and AC marked
and joined to make the smaller inner
triangle as above

   (e) Isosceles

C2 (a) Points P, Q and R plotted and joined
as above

   (b) A scalene triangle has three sides of
different length.

   (c) Mid-point of each edge marked and
joined as above

   (d) The pupil's comments (see teacher's
notes above)

C3 (a) The pupil's scalene triangle with the
mid-points marked and joined

   (b) The area of each smaller triangle is $\frac{1}{4}$
of the area of the larger one.

C4

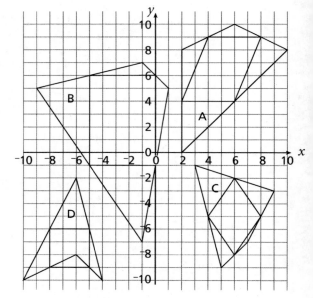

A: Parallelogram

B: Rectangle

C: Rhombus

D: Square

C5 The pupil's investigation (see teacher's
notes above)

D Mid-points and coordinates (p 225)

D1 (a) (5, ⁻2)   (b) (4, 3)   (c) (⁻3, 1½)

D2 (a) The line segment from (0, 1) to
(8, 1) drawn on a grid, mid-point is
(4, 1)

   (b) The pupil's horizontal line segments
with the mid-points marked

   (c) The pupil's table

   (d) The pupil's description of the rule
that the mid-point of the line
segment from (a, c) to (b, c) is
$\left(\frac{a + b}{2}, c\right)$

D3 (a) The pupil's vertical line segments with the mid-points marked

(b) The pupil's description of the rule that the mid-point of the line segment from (c, a) to (c, b) is $\left(c, \dfrac{a + b}{2}\right)$

D4 The pupil's investigation and description of the rule that the mid-point of the line segment from (a, c) to (b, d) is $\left(\dfrac{a + b}{2}, \dfrac{a + b}{2}\right)$

D5 (a) $(5, 3)$ (b) $(6, 5)$ (c) $(^-1, ^-2)$

(d) $(^-3, 5\frac{1}{2})$ (e) $(3, 6)$ (f) $(1\frac{1}{2}, ^-2)$

E Coordinate code (p 226)

E1 The points are $(1, 1)$, $(2, ^-1)$, $(^-2, 0)$ and $(2, 2)$ giving the letters ITKE and the word KITE

E2 The points are $(2, ^-1)$, $(1, 0)$, $(1, 1)$, $(^-2, ^-1)$ and $(2, 0)$ giving the letters TNIPO and the word POINT

E3 The pupil's problems that lead to the points $(1, ^-1)$, $(^-1, ^-1)$, $(^-2, ^-2)$, $(^-2, 2)$, $(0, ^-1)$ and $(2, 2)$ in some order

Curve stitching

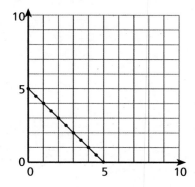

What progress have you made? (p 227)

1   $5 \to \mathbf{17}$

   $^-1 \to \mathbf{^-7}$

   $\mathbf{2 \to 5}$

2 $y = 14$

3 (a)–(c)

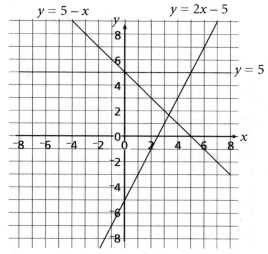

4 (a) $(5, 5)$ (b) $(13, ^-2)$

5 The diagonals of the parallelogram cross at $(\frac{1}{2}, 0)$.

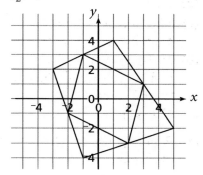

Practice booklet

Sections A and B (p 101)

1 (a) (i) $4 \to \mathbf{12}$ (ii) $5 \to \mathbf{11}$

          $2 \to \mathbf{8}$ $2 \to \mathbf{2}$

          $0 \to \mathbf{4}$ $^-1 \to \mathbf{^-7}$

         $^-2 \to \mathbf{0}$ $3 \to \mathbf{5}$

    (iii) $1 \to \mathbf{8}$ (iv) $3 \to \mathbf{6}$

          $5 \to \mathbf{0}$ $0 \to \mathbf{5}$

          $0 \to \mathbf{10}$ $6 \to \mathbf{7}$

         $^-1 \to \mathbf{12}$ $^-3 \to \mathbf{4}$

(b)

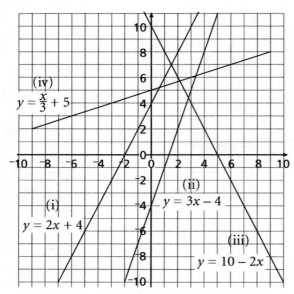

2 (a) $n \rightarrow n + 3$ (b) $n \rightarrow 4n$

   (c) $n \rightarrow n^2 - 2$ (d) $n \rightarrow 3n - 1$

Sections C, D and E (p 102)

1 (a) $(4, 4)$ (b) $(3, 10\frac{1}{2})$ (c) $(3, 6)$

   (d) $(2, {}^-7)$ (e) $(3, 6)$ (f) $(5\frac{1}{2}, 6\frac{1}{2})$

   (g) $(-\frac{1}{2}, 2\frac{1}{2})$ (h) $(4, 1)$ (i) $(-\frac{1}{2}, {}^-3\frac{1}{2})$

2 (a) Polygon ABCDEF as shown below.

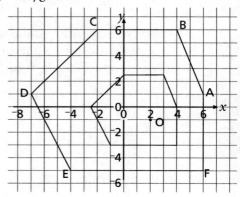

(b) Point O plotted with the points of the smaller polygon as above; the pupil's comments such as 'The small polygon is a scaled-down copy of the larger.'

Review 4 (p 228)

1 (a)
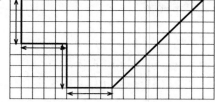

2

(b)

2

2 34.6 cm

3 (a)

```
1 | 2 5 8
2 | 1 1 2 2 4 5 8
3 | 0 1 2 4 4 5 6 8
4 | 0 1
```

4

2   (b) $41 - 12 = 29$ (c) 29 cm

1   **4** $280 = 2^3 \times 5 \times 7$

5 (a) 300   (b) 4000   (c) 50

8   (d) 0.3   (e) 0.03

6 (a) $x = 2$ (b) $p = 7$ (c) $z = 8$

6   (d) $d = 9$ (e) $r = 11$ (f) $m = 5$

7 (a) Scale factor 3

3   (b) $a = 2$ $b = 4.5$

3   **8** (a) 600 000   (b) 3200   (c) 0.542

9 (a) 0.08   (b) 12

4   (c) 35   (d) 0.015

10 (a) 1 cm represents 4 metres.

(b) 20 metres

3   (c) 12.5 cm

11 (a) 30   (b) 0.06

4   (c) 0.04   (d) 0.9

12 (a) About $80 \times 40 = 3200 \, \text{m}^2$

2   (b) About $40 \div 0.8 = 50 \, \text{m}$

13 (a) $3p + 7$

1   (b) $5p + 1$

2   (c) $3p + 7 = 5p + 1$ $p = 3$

1   (d) Perimeter = 16

2 **14** (a) $\dfrac{1300}{8} = 162.5 \, \text{g}$ (b)   32 g

(c) For example, the mandarins are
heavier than the clementines.
The weight of the mandarins is more
2   spread out than the weight of the
clementines.

15 If n is the number they started with,
2   $6n - 14 = 5(n + 1)$ $n = 19$

16 $2^{10} = 1024$ $6^4 = 1296$
2   6^4 is larger by 272.

17 (a) The pupil's accurate drawing
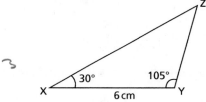

3

1   (b) 4.2 cm

(c) The pupil's ruler and compasses
1   construction of the perpendicular
from Y to XZ

18 (a) 5130 yen

2   (b) £47.37 to nearest penny.

19 (a) $50 \times 0.4 = 20$

(b) $800 \times 20 = 16\,000$

(c) $300 \times 0.7 = 210$

4   (d) $0.04 \times 30 = 1.2$

20

Time (seconds)	Frequency
10–15	2
15–20	8
20–25	11
25–30	3

The modal interval was 20–25 seconds.

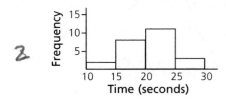

21 (a) a^9 (b) a^5 (c) a^2

22 (a) The pupil's accurate drawing

(b) (ii) 033° (ii) 122°

(c) X marked accurately on the pupil's diagram.

(d) 1780 m = 2000 m to 1 s.f.

23 (a) RICE (b) FISH

(c) MEAT (d) SOUP

(e) CAKE

Mixed questions 4 (Practice booklet p 102)

1 (a) 2 (b) 1.5 (c) $\frac{1}{3}$

(d) 4 (e) $\frac{1}{2}$

2 (a) 300 (b) $2 \times 3^2 \times 5^2$

3 $a = 3$ cm, $b = 22.5$ cm

4 (a) $r = 2$ (b) $a = 7$

(c) $z = 5$ (d) $t = 9$

5 (a) 34 km (b) 5.6 cm

6 $x = 7$

7 385 g

8 (a) The pupil's accurate diagram

(b) 7.6 km

(c) 235°

9 (a) 1.8 m^2 (b) 70

10 (a) (i) About A$14

(ii) About £14.50

(b) £12 is about A$33, so the A$31 in Australia is cheaper.

11 (a) $\frac{3}{8}$ (b) $1\frac{1}{2}$ (c) $\frac{4}{6} = \frac{2}{3}$

(d) $\frac{15}{4} = 3\frac{3}{4}$

12 (a) 18 (b) 24 (c) 25

13 (a) 14 (b) 0.012 (c) 50

(d) 0.2

14 (a) c^8 (b) d^6 (c) e^5

15 A: mean £6.35k range £8.5k

B: mean £7.2k range £3.1k

Showroom A has a lower mean but a larger range of prices.

Showroom B has a higher mean but less variation in prices.

16 (a) The pupil's accurate drawing of triangle RST

(b) Angle bisectors of R and S constructed

(c) 2.9 cm

17 (a) 1 : 5000 (b) 1 : 2500

(c) 1 : 500 000

18 The pupil's scatter graph